Authored by B. V. Senthil Kumar and Hemen Dutta

Discrete Mathematical Structures

Mathematics and Its Applications: Modelling, Engineering, and Social Sciences

Series Editor: Hemen Dutta

Discrete Mathematical Structures
A Succinct Foundation
B. V. Senthil Kumar and Hemen Dutta

Concise Introduction to Logic and Set Theory
Iqbal H. Jebril and Hemen Dutta

Tensor Calculus and Applications
Simplified Tools and Techniques
Bhaben Kalita

For more information about this series, please visit:
www.crcpress.com/Mathematics-and-its-applications/book-series/MES

Authored by B. V. Senthil Kumar and Hemen Dutta

Discrete Mathematical Structures

A Succinct Foundation

CRC Press
Taylor & Francis Group
Boca Raton London New York

CRC Press is an imprint of the
Taylor & Francis Group, an **informa** business

CRC Press

Taylor & Francis Group

6000 Broken Sound Parkway NW, Suite 300

Boca Raton, FL 33487-2742

CRC Press is an imprint of Taylor & Francis Group, an Informa business

No claim to original U.S. Government works

Printed on acid-free paper

International Standard Book Number-13: 978-0-367-14869-0 (Hardback)

This book contains information obtained from authentic and highly regarded sources. Reasonable efforts have been made to publish reliable data and information, but the author and publisher cannot assume responsibility for the validity of all materials or the consequences of their use. The authors and publishers have attempted to trace the copyright holders of all material reproduced in this publication and apologize to copyright holders if permission to publish in this form has not been obtained. If any copyright material has not been acknowledged, please write and let us know so we may rectify in any future reprint.

Library of Congress Cataloging-in-Publication Data

Names: B. V. Senthil Kumar, author. | Dutta, Hemen, 1981- author.
Title: Discrete mathematical structures : a succinct foundation / by B.V.
Senthil Kumar and Hemen Dutta.
Description: Boca Raton, FL : CRC Press/Taylor & Francis Group, 2020. |
Series: Mathematics and its applications : modelling, engineering, and
social sciences
Identifiers: LCCN 2019009359 | ISBN 9780367148690 (hardback : alk. paper) |
ISBN 9780429053689 (ebook)
Subjects: LCSH: Discrete mathematics
Classification: LCC QA297.4 .K86 2020 | DDC 518/.25–dc23
LC record available at https://lccn.loc.gov/201900935

Visit the Taylor & Francis Web site at
http://www.taylorandfrancis.com

and the CRC Press Web site at
http://www.crcpress.com

Printed and bound in Great Britain by
TJ International Ltd, Padstow, Cornwall

Contents

Preface

The aim of this book is to provide a concise introduction to some significant topics covered in the subject Discrete Mathematics. Various themes of discrete mathematics have extensive applications in several courses in disciplines like Computer Science, Engineering, and Information Technology. This motivated us to provide a resource in the form of a textbook for undergraduate and postgraduate students adopting relevant courses in Computer Science, Mathematics, Engineering, Information Technology, etc. While in some courses this book can be adopted as a text book, there are several other courses where this book will serve the need of a reference book. This book also will serve as a Handbook for the teachers in class room teaching as it contains many solved problems and problems for practice which are included at the end of each subtopic. In order to meet the needs of the learners, we have included essential topics in this book, such as Logics and Proofs, Combinatorics, Graphs, Algebraic Structures, Lattices and Boolean Algebra.

In **Chapter 1: Logics and Proofs**, vital concepts from basic level of logic and proofs are provided. In **Chapter 2: Combinatorics**, we have dealt with the idea of mathematical induction with more number of solved problems for better understanding of the reader. Also, various other topics in combinatorics like pigeonhole principle, permutations and combinations, and recurrence relation are discussed in meticulous approach. In **Chapter 3: Graphs**, fundamental notion of graph theory and various types of graphs are introduced with lot of illustrations. In **Chapter 4: Algebraic Structures**, the elementary tools of discrete mathematics such as algebraic structure, semigroup, monoid, abelian group, subgroup, cosets, Lagrange's theorem, normal subgroup, homomorphism of groups, rings and fields are provided with sufficient theorems with proofs and examples. In **Chapter 5: Lattices and Boolean Algebra**, the concepts of partially ordered sets, lattices, Boolean algebra and their properties are focussed. These concepts are useful in many different types of computational circuits.

We thank all the authors who have contributed extensively in the field of Discrete Mathematics and our family members and friends who have motivated us to bring our objective in the form of a textbook. We are indebted to the Management of Nizwa College of Technology, Nizwa, Oman for their constant support and encouragement during the preparation of this book. We would like to thank the editors in charge for this book project and supporting staff at CRC Press, Taylor & Francis Group, for their timely cooperation in publishing this book. We also welcome productive suggestions and comments to improve the quality of the book for next edition.

B. V. Senthil Kumar, Nizwa, Oman
Hemen Dutta, Guwahati, India

Authors

B. V. Senthil Kumar is serving in the Section of Mathematics, Department of Information Technology, Nizwa College of Technology, Nizwa, Oman. His areas of interest are solution and stability of functional, differential, and difference equations; operations research, statistics; and discrete mathematics. He obtained his Ph.D. Degree in 2015 and has 18 years of teaching and research experience. He has published more than 40 research papers in National and International reputed journals. He has authored the books titled *Functional Equations and Inequalities*: *Solution and Stability Results* (published by World Scientific Publishing Company) and *Probability & Queueing Theory* (published by KKS Publishers, Chennai, India) to his credit. He has also contributed some book chapters. He has delivered invited talks in various institutions and also organised many academic and non-academic events. He is a member of many mathematical societies.

Hemen Dutta is a faculty member at the Department of Mathematics, Gauhati University, India. He did his Master of Science in Mathematics, Post Graduate Diploma in Computer Application, and Ph.D. in Mathematics from Gauhati University, India. He received his M.Phil in Mathematics from Madurai Kamaraj University, India. He currently teaches subjects like real analysis, functional analysis, algebra, mathematical logic, computer applications, etc. His primary research interest includes areas of mathematical analysis. He has to his credit several research papers, some book chapters, and few books. He has delivered talks at different institutions and organised a number of academic events. He is a member of several mathematical societies.

1

Logics and Proofs

1.1 Introduction

In this chapter, we discuss propositional logic and various methods of proving validity of propositions. The concept of logic has many applications in computer science to develop computer programs, to verify the logic of program and also in electronics to design circuits.

1.2 Proposition

A proposition (or statement) is a declarative sentence which is true or false, but not both. Consider, for example,

(i) The year 2000 is a leap year.

(ii) $5 + 3 = 7$.

(iii) $x = 1$ is a solution of $x^3 = 1$.

(iv) Close the door.

In the above, (i)–(iii) are propositions, whereas (iv) is not a proposition. Moreover, (i) and (iii) are true, while (ii) is false.

1.3 Compound Propositions

Many propositions are composite, that is, composed of subpropositions and various connectives discussed in the next section. Such composite propositions are called compound propositions. A proposition is said to be primitive if it cannot be broken into smaller propositions, that is, if it is not composite.

Examples:

(i) Apples are red, and milk is white.

(ii) Jack is brilliant or is a hardworking student.

1

Note:

The truth value of a compound proposition is obtained by the truth values of its subpropositions together with the way in which they are connected to form the compound propositions.

1.4   Truth Table

A truth table lists all possible combinations of truth values of the propositions in the left most column and the truth values of the resulting propositions in the right most column.

1.5   Logical Operators

1.5.1   Negation

If P is a statement, then negation of P written as $\neg P$ or $\sim P$ is read as "Not P". The truth table for the operator "negation" is shown below.

Negation

P	$\neg P$
T	F
F	T

Example:

P: Apple is red.

$\neg P$: Apple is not red.

1.5.2   Conjunction

The conjunction of two statements P and Q is the statement $P \wedge Q$ which is read as "P and Q". The statement $P \wedge Q$ has a truth value T whenever both P and Q have the truth value T; otherwise, it has a truth value F. The conjunction is defined by the truth table below.

Conjunction

P	Q	$P \wedge Q$
T	T	T
T	F	F
F	T	F
F	F	F

Example:

P: John worked hard.

Q: John passed the examination.

$P \wedge Q$: John worked hard, and he passed the examination.

1.5.3   Disjunction

The disjunction of two statements P and Q is the statement $P \vee Q$ and has a truth value F only when both P and Q have truth value F; otherwise, it has a truth value T. The disjunction is defined by the truth table shown below.

Disjunction

P	Q	$P \vee Q$
T	T	T
T	F	T
F	T	T
F	F	F

Example:

P: $2 + 4 = 6$ (T).

Q: $2 > 10$ (F).

$P \vee Q$: $2 + 4 = 6$ or $2 > 10$ is true.

1.5.4   Molecular Statements

The statements that contain one or more atomic statements and some connectives are called molecular statements.

Examples: $\neg P$, $P \wedge \neg Q$, $\neg P \vee \neg Q$, etc.

1.5.5   Conditional Statement [If . . . then] [$\rightarrow$]

If P and Q are any two statements, then the statement $P \rightarrow Q$ which is read as "If P then Q" is called a conditional statement. Here, P is called "antecedent", and Q is called "consequent". The truth table is shown below.

If . . . then

P	Q	$P \rightarrow Q$
T	T	T
T	F	F
F	T	T
F	F	T

Note:

$P \rightarrow Q$ has a truth value F if P has the truth value T and Q has the truth value F. In all the remaining cases, it has the truth value T.

Example:

P: It is hot.

Q: $2 + 3 = 5$.

$P \rightarrow Q$: If it is hot, then $2 + 3 = 5$.

1.5.6 Biconditional [If and only if or iff] [$\leftrightarrow$ or $\rightleftharpoons$]

If P and Q are any two statements, then the statement $P \leftrightarrow$ or $P \rightleftharpoons Q$ which is read as "P if and only if Q" is called biconditional statement. The statement $P \leftrightarrow Q$ has the truth value T whenever both P and Q have identical truth values. The truth table is shown below.

If and only if

P	Q	$P \leftrightarrow Q$
T	T	T
T	F	F
F	T	F
F	F	T

Example:

P: John is rich.

Q: John is happy.

$P \leftrightarrow Q$: John is rich if and only if he is happy.

1.5.7 Solved Problems

1. Give the contrapositive statement of the statement "If there is rain, then I buy an umbrella".

   Solution.

   Let P: "There is rain" and Q: "I buy an umbrella".

   Then the given statement is $P \rightarrow Q$.

   Its contrapositive is $\neg Q \rightarrow \neg P$.

2. Construct the truth table for $P \rightarrow \neg Q$.

   Solution.

   The truth table is shown below.

Truth Table for $P \rightarrow \neg Q$

P	Q	$\neg Q$	$P \rightarrow \neg Q$
T	T	F	F
T	F	T	T
F	T	F	T
F	F	T	T

3. Find the truth table for $P \rightarrow Q$.

    Solution.
    The truth table is shown below.

    Truth Table for $P \rightarrow Q$

P	Q	$P \rightarrow Q$
T	T	T
T	F	F
F	T	T
F	F	T

4. Construct the truth table for the compound proposition
    $(P \rightarrow Q) \leftrightarrow (\neg P \rightarrow \neg Q)$.

    Solution.
    The truth table is shown below.

    Truth Table for $(P \rightarrow Q) \leftrightarrow (\neg P \rightarrow \neg Q)$

P	Q	$P \rightarrow Q$	$\neg P$	$\neg Q$	$\neg P \rightarrow \neg Q$	$(P \rightarrow Q) \leftrightarrow (\neg P \rightarrow \neg Q)$
T	T	T	F	F	T	T
T	F	F	F	T	T	F
F	T	T	T	F	F	F
F	F	T	T	T	T	T

5. What are the contrapositive, the converse, and the inverse of the
    following conditional statement?
    "If you work hard, then you will be rewarded".

    Solution.
    P: You work hard.
    Q: You will be rewarded.
    $\neg P$: You will not work hard.
    $\neg Q$: You will not be rewarded.
    Converse: $Q \rightarrow P$: If you will be rewarded, then you work hard.
    Contrapositive: $\neg Q \rightarrow \neg P$: If you will not be rewarded, then you
    will not work hard.
    Inverse: $\neg P \rightarrow \neg Q$: If you will not work hard, then you will not
    be rewarded.

6. Construct a truth table for the compound proposition
    $(P \rightarrow Q) \rightarrow (Q \rightarrow P)$.

    Solution.
    The truth table is shown below.

Truth Table for $(P \to Q) \to (Q \to P)$

P	Q	$P \to Q$	$Q \to P$	$(P \to Q) \to (Q \to P)$
T	T	T	T	T
T	F	F	T	T
F	T	T	F	F
F	F	T	T	T

1.5.8   Tautology

A statement formula which is **true** regardless of the truth values of the statements which replace the variables in it is called a **tautology** or a universally valid formula or a logical truth.

Example: $P \vee \neg P$ is a tautology.

1.5.9   Contradiction

A statement formula which is **false** regardless of the truth values of the statements which replace variables in it is called a **contradiction**.

Example: $P \wedge \neg P$ is a contradiction.

1.5.10   Contingency

A statement formula which is neither tautology nor contradiction is called **contingency**.

Example: $P \to Q$ is a contingency.

Note: To determine whether a given formula is a tautology or a contradiction, construct the truth table. But this process is very lengthy since the truth table will have 2^n rows for n statements.

1.5.11   Equivalence Formulas

Let A and B be two statement formulas, and let $p_1, p_2, \ldots, p_n$ denote all the variables occurring in both A and B. If the truth value of A is equal to the truth value of B for every one of the 2^n possible sets of truth values assigned to $p_1, p_2, \ldots, p_n$, then A and B are said to be equivalent.

Assuming that the variables and assignment of truth values to the variables appear in the same order in the truth tables of A and B, the final columns in the truth tables for A and B are identical if A and B are equivalent.

Examples:

(i) $\neg\neg P$ is equivalent to P.

(ii) $P \vee \neg P$ is equivalent to $Q \vee \neg Q$.

Remark: We know that $A \rightleftharpoons B$ is true whenever A and B have identical truth values. This means A is equivalent to B ($\Leftrightarrow$) if and only if $A \rightleftharpoons B$ is a tautology.

1.5.12 Equivalent Formulas

(1) $P \vee \neg P$, $P \wedge \neg P$ [**Idempotent laws**]

(2) $(P \vee Q) \vee R \Leftrightarrow P \vee (Q \vee R)$, $(P \wedge Q) \wedge R \Leftrightarrow P \wedge (Q \wedge R)$
   [**Associative laws**]

(3) $P \vee Q \Leftrightarrow Q \vee P$, $P \wedge Q \Leftrightarrow Q \wedge P$ [**Commutative laws**]

(4) $P \vee (Q \wedge R) \Leftrightarrow (P \vee Q) \wedge (P \vee R)$, $P \wedge (Q \vee R) \Leftrightarrow (P \wedge Q) \vee (P \wedge R)$
   [**Distributive laws**]

(5) $\neg(P \vee Q) \Leftrightarrow \neg P \wedge \neg Q$, $\neg(P \wedge Q) \Leftrightarrow \neg P \vee \neg Q$
   [**De Morgan's laws**]

1.5.13 Duality Law

Two formulas A and $A^\star$ are duals of each other if either one can be obtained from the other by replacing $\wedge$ by $\vee$ and $\vee$ by $\wedge$. The connectives $\vee$ and $\wedge$ are also called duals of each other.

Examples:
Write the duals of (i)(P $\vee$ Q) $\wedge$ R,   (ii)(P $\wedge$ Q) $\vee$ T.
The duals are (i)(P $\wedge$ Q) $\vee$ R,   (ii)(P $\vee$ Q) $\wedge$ T.

1.5.14 Tautological Implication

A statement A is said to tautologically imply a statement B if $A \rightarrow B$ is a tautology. We use the notation $\Rightarrow$.

Remark: To prove $P \Rightarrow Q$, we assume P to be true and prove Q to be true. Otherwise, assume Q to be false, and prove P to be false also. Construction of truth table is another method for proving the implication.

1.5.15 Some More Equivalence Formulas

1. $P \wedge \neg P \Leftrightarrow F$, $P \vee \neg P \Leftrightarrow T$ [**Complement laws**]

2. $P \vee T \Leftrightarrow T$, $P \wedge F \Leftrightarrow F$ [**Dominance laws**]

3. $P \wedge T \Leftrightarrow P$, $P \vee F \Leftrightarrow P$ [**Identity laws**]

4. $P \vee (P \wedge Q) \Leftrightarrow P$, $P \wedge (P \vee Q) \Leftrightarrow P$ [**Absorption laws**]

5. $\neg(\neg P) \Leftrightarrow P$ [**Double Negation law**]

6. $P \rightarrow Q \Leftrightarrow \neg Q \rightarrow \neg P$ [**Contrapositive law**]

7. $P \rightarrow Q \Leftrightarrow \neg P \vee Q$ [**Conditional as disjunction**]

8. $P \rightleftharpoons Q \Leftrightarrow (P \rightarrow Q) \wedge (Q \rightarrow P)$ [**Biconditional as conjunction**]

1.5.16   Solved Problems

1. Using truth table, show that the proposition $P \vee \neg(P \wedge Q)$ is a tautology.

   Solution.

   The truth table is shown below. Since all the entries in the last column are T, the given proposition is a tautology.

   Truth Table of $P \vee \neg(P \wedge Q)$

P	Q	$P \wedge Q$	$\neg(P \wedge Q)$	$P \vee \neg(P \wedge Q)$
T	T	T	F	T
T	F	F	T	T
F	T	F	T	T
F	F	F	T	T

2. Express $A \leftrightarrow B$ in terms of the connectives $\{\wedge, \neg\}$.

   Solution.
   $$A \leftrightarrow \Leftrightarrow (A \rightarrow B) \wedge (B \rightarrow A)$$
   $$\Leftrightarrow (\neg A \vee B) \wedge (\neg B \vee A).$$

3. Show that $(p \rightarrow r) \wedge (q \rightarrow r)$ and $(p \vee q) \rightarrow r$ are logically equivalent.

   Solution.
   $$(p \rightarrow r) \wedge (q \rightarrow r)$$
   $$\Leftrightarrow (\neg p \vee r) \wedge (\neg q \vee r) \quad \text{(conditional as disjunction)}$$
   $$\Leftrightarrow (\neg p \wedge \neg q) \vee r \quad \text{(Distributive law)}$$
   $$\Leftrightarrow \neg(p \vee q) \vee r \quad \text{(De Morgan's law)}$$
   $$\Leftrightarrow (p \vee q) \rightarrow r \quad \text{(conditional as disjunction)}.$$

4. Is $(\neg p \wedge (P \vee q)) \rightarrow q$ is a tautology.

   Solution.
   $$(\neg p \wedge (p \vee q)) \rightarrow q$$
   $$\Leftrightarrow (\neg p \wedge p) \vee (\neg p \wedge q) \rightarrow q \quad \text{(Distributive law)}$$
   $$\Leftrightarrow F \vee (\neg p \wedge q) \rightarrow q \qquad [p \wedge \neg p \Leftrightarrow F]$$
   $$\Leftrightarrow (\neg p \wedge q) \rightarrow q \qquad [p \vee F \Leftrightarrow p]$$
   $$\Leftrightarrow \neg(\neg p \wedge q) \vee q \qquad [p \rightarrow q \Leftrightarrow \neg p \vee q]$$
   $$\Leftrightarrow (p \vee \neg q) \vee q \qquad \text{(De Morgan's law)}$$
   $$\Leftrightarrow p \vee q \vee \neg q \qquad \text{(Associative law)}$$
   $$\Leftrightarrow p \vee T \qquad [p \vee \neg p \Leftrightarrow T]$$
   $$\Leftrightarrow T.$$

   Therefore, the given statement is a tautology.

5. Show that the propositions $p \to q$ and $\neg p \lor q$ are logically equivalent.

Solution.

From the truth table, $p \to q$ and $\neg p \lor q$ are equivalent.

Truth Table of $p \to q$ and $\neg p \lor q$

p	q	$\neg p$	$p \to q$	$\neg p \lor q$
T	T	F	T	T
T	F	F	F	F
F	T	T	T	T
F	F	T	T	T

1.6   Normal Forms

1.6.1   Principal Disjunctive Normal Form or Sum of Products Canonical Form

Consider two statements P and Q. Consider a possible formula using conjunction as follows: $P \land Q$, $\neg P \land Q$, $P \land \neg Q$, $\neg P \land \neg Q$ (Duplication is not allowed and only distinct formulas are considered). We call the above terms as **"minterms"**.

For a given formula, an equivalent formula consisting of disjunction of minterms only is known as **Principal Disjunctive Normal Form** (PDNF) or **Sum of Products Canonical Form**.

Procedure I:

For every truth value T in the truth table of the given formula, select the minterm which also has the value T for the same combination of the truth values of P and Q. The disjunction of these minterms will then be equivalent to the given formula. From the table below, we observe that

Truth Table showing Disjunctions of P and Q

P	Q	$P \land Q$	$\neg P$	$\neg Q$	$\neg P \land Q$	$P \land \neg Q$	$\neg P \land \neg Q$	$P \to Q$	$P \lor Q$
T	T	T	F	F	F	F	F	T	T
T	F	F	F	T	F	T	F	F	T
F	T	F	T	F	T	F	F	T	T
F	F	F	T	T	F	F	T	T	F

$$P \to Q \Leftrightarrow (P \land Q) \lor (\neg P \land Q) \lor (\neg P \land \neg Q)$$
$$P \lor Q \Leftrightarrow (P \land Q) \lor (P \land \neg Q) \lor (\neg P \land Q).$$

Procedure II:

This is explained in the following example:

$$P \vee Q \Leftrightarrow \neg P \wedge (Q \vee \neg Q) \vee (Q \wedge (P \vee \neg P)) \quad [\text{since } A \wedge T \Leftrightarrow A]$$
$$\Leftrightarrow (\neg P \wedge Q) \vee (\neg P \wedge \neg Q) \vee (Q \wedge P) \vee (Q \wedge \neg P) \quad [\text{Distributive laws}]$$
$$\Leftrightarrow (\neg P \wedge Q) \vee (\neg P \wedge \neg Q) \vee (P \wedge Q) \quad [P \vee P \Leftrightarrow P]$$

Note:

1. The number of minterms appearing in the normal form is the same as the number of entries with the truth value T in the truth table of the given formula. Thus, every formula which is not a contradiction has an equivalent PDNF.

2. If a formula is a tautology, then all the minterms will appear in its PDNF.

3. To show that two formulas are equivalent, obtain PDNFs of the two formulas. If the normal forms are identical, then both the formulas are equivalent.

4. Minterms of three variables are $P \wedge Q \wedge R$, $P \wedge Q \wedge \neg R$, $P \wedge \neg Q \wedge R$, $\neg P \wedge Q \wedge R$, $P \wedge \neg Q \wedge \neg R$, $\neg P \wedge Q \wedge \neg R$, $\neg P \wedge \neg Q \wedge \neg R$, $\neg P \wedge \neg Q \wedge \neg R$.

1.6.2 Principal Conjunctive Normal Form or Product of Sum Canonical Form

For a given number of variables, the "maxterms" consist of disjunctions in which each variable or its negation, but not both, appear only once. For a given formula, an equivalent formula consisting of conjunctions of maxterms only is known as its **Principal Conjunctive Normal Form (PCNF)** or **Product of Sum of Canonical Form**.

Note:

If the PDNF or PCNF of a given formula A consisting of n variables is known, then the PDNF or PCNF of $\neg A$ will consist of the disjunction (or conjunction) of the remaining minterms (or maxterms) which do not appear in the PDNF or PCNF of A. From $A \Leftrightarrow \neg\neg A$, one can obtain the PDNF or PCNF of A by repeated applications of De Morgan's laws to the PDNF or PCNF of $\neg A$.

1.6.3 Solved Problems

1. Obtain the PDNF and PCNF of $(P \wedge Q) \vee (\neg P \wedge R)$.

   Solution.

   $$(P \wedge Q) \vee (\neg P \wedge R)$$
   $$\Leftrightarrow ((P \wedge Q) \wedge T) \vee ((\neg P \wedge R) \wedge T) \quad [\text{since } A \wedge T \Leftrightarrow A]$$
   $$\Leftrightarrow ((P \wedge Q) \wedge (R \vee \neg R)) \vee ((\neg P \wedge R) \wedge (Q \vee \neg Q))$$
   $$\qquad\qquad\qquad\qquad\qquad\qquad [\text{since } P \vee \neg P \Leftrightarrow T]$$

$\Leftrightarrow (P \wedge Q \wedge R) \vee (P \wedge Q \wedge \neg R) \vee (\neg P \wedge Q \wedge R) \vee (\neg P \wedge \neg Q \wedge R)$

[Distributive laws]

is the required PDNF.

The remaining minterms are

$$P \wedge \neg Q \wedge R, \quad P \wedge \neg Q \wedge \neg R, \quad \neg P \wedge Q \wedge \neg R, \quad \neg P \wedge \neg Q \wedge \neg R.$$

The required PCNF is

$\neg ((P \wedge \neg Q \wedge R) \vee (P \wedge \neg Q \wedge \neg R) \vee (\neg P \wedge Q \wedge \neg R)$
$\quad \vee (\neg P \wedge \neg Q \wedge \neg R))$
$\Leftrightarrow (\neg P \vee Q \vee \neg R) \wedge (\neg P \vee Q \vee R) \wedge (P \vee \neg Q \vee R) \wedge (P \vee Q \vee R).$

2. Obtain the PCNF and PDNF of $(\neg P \rightarrow R) \wedge (Q \leftrightarrow P)$ by using equivalences.

Solution.
$\quad (\neg P \rightarrow R) \wedge (Q \leftrightarrow P)$
$\Leftrightarrow (P \vee R) \wedge ((Q \rightarrow P) \wedge (P \rightarrow Q))$
$\qquad$ [since $P \rightarrow Q \Leftrightarrow \neg P \vee Q, P \leftrightarrow Q \Leftrightarrow (P \rightarrow Q) \wedge (Q \rightarrow P)$]
$\Leftrightarrow (P \vee R) \wedge ((\neg Q \vee P) \wedge (\neg P \vee Q))$
$\Leftrightarrow ((P \vee R) \vee (Q \wedge \neg Q)) \wedge ((\neg Q \vee P) \vee (R \vee \neg R)) \wedge ((\neg P \vee Q)$
$\qquad \vee (R \wedge \neg R))$
$\Leftrightarrow (P \vee Q \vee R) \wedge (P \vee \neg Q \vee R) \wedge (P \vee \neg Q \vee R) \wedge (P \vee \neg Q \vee \neg R)$
$\qquad \wedge (\neg P \vee Q \vee R) \wedge (\neg P \vee Q \wedge \neg R)$
$\Leftrightarrow (P \vee Q \vee R) \wedge (P \vee \neg Q \vee R) \wedge (P \vee \neg Q \vee \neg R) \wedge (\neg P \vee Q \vee R)$
$\qquad \wedge (\neg P \vee Q \vee \neg R)$

which is the required PDNF.

The remaining minterms are

$$P \vee Q \vee \neg R, \quad \neg P \vee \neg Q \vee R, \quad \neg P \vee \neg Q \vee \neg R.$$

The required PCNF is

$\qquad \neg ((P \vee Q \vee \neg R) \wedge (\neg P \vee \neg Q \vee R) \wedge (\neg P \vee \neg Q \vee \neg R))$
$\Leftrightarrow (\neg P \wedge \neg Q \wedge R) \vee (P \wedge Q \neg R) \vee (P \wedge Q \wedge R).$

3. Find the PDNF of $(Q \vee (P \wedge R)) \wedge \neg ((P \vee R) \wedge Q)$.

Solution.

$\quad (Q \vee (P \wedge R)) \wedge \neg ((P \vee R) \wedge Q)$
$\Leftrightarrow (Q \vee (P \wedge R)) \wedge ((\neg P \wedge \neg R) \vee \neg Q)$ [by De Morgan's laws]
$\Leftrightarrow Q \wedge (\neg P \wedge \neg R) \vee (Q \wedge \neg Q) \vee (P \wedge R \wedge \neg P \wedge \neg R) \vee (P \wedge R \wedge \neg Q)$

[using Distributive law]

$$\Leftrightarrow (\neg P \wedge Q \wedge \neg R) \vee F \vee (F \wedge R) \vee (P \wedge \neg Q \wedge R)$$
$$\Leftrightarrow (\neg P \wedge Q \wedge \neg R) \vee (P \wedge \neg Q \wedge R).$$

4. Prove that $((P \vee Q) \wedge \neg(\neg P \wedge (\neg Q \vee \neg R)))) \vee (\neg P \wedge \neg Q) \vee (\neg P \wedge \neg R)$ is a tautology.

    Solution.
    Consider

    $$\neg(\neg P \wedge (\neg Q \vee \neg R)) \Leftrightarrow P \vee \neg(\neg Q \vee \neg R) \ \text{[De Morgan's law]}$$
    $$\Leftrightarrow P \vee (Q \wedge R) \ \text{[De Morgan's law]}$$
    $$\Leftrightarrow (P \vee Q) \wedge (P \vee R) \ \text{[De Morgan's law]}. \ (1.1)$$

    Now, consider

    $$(\neg P \wedge \neg Q) \vee (\neg P \wedge \neg R)$$
    $$\Leftrightarrow (\neg(P \vee Q) \vee \neg(P \vee R) \quad \text{[De Morgan's law]}$$
    $$\Leftrightarrow \neg((P \vee Q) \wedge (P \vee R)) \quad \text{[De Morgan's law]}. \qquad (1.2)$$

    From (1.1) and (1.2), we obtain

    $$((P \vee Q) \wedge (P \vee Q) \wedge (P \vee R)) \vee \neg((P \vee Q) \wedge (P \vee R))$$
    $$\Leftrightarrow ((P \vee Q) \wedge (P \vee R)) \vee \neg((P \vee Q) \wedge (P \vee R))$$
    $$\Leftrightarrow T.$$

    Hence, the given statement formula is a tautology.

5. Prove that $(P \to Q) \wedge (Q \to R) \Rightarrow (P \to R)$.

    Solution.
    It is enough to prove $(P \to Q) \wedge (Q \to R) \Rightarrow (P \to R)$ is a tautology.
    Let $S = (P \to Q) \wedge (Q \to R) \Rightarrow (P \to R)$.

    Since the last column is T for all eight combinations as shown in the table below, the given statement formula is a tautology.

Truth Table showing $(P \to Q) \wedge (Q \to R) \Rightarrow (P \to R)$

P	Q	R	$P \to Q$	$Q \to R$	$P \to R$	$(P \to Q) \wedge (Q \to R)$	S
T	T	T	T	T	T	T	T
T	T	F	T	F	F	F	T
T	F	T	F	T	T	F	T
F	T	T	T	T	T	T	T
T	F	F	F	T	F	F	T
F	T	F	T	F	T	F	T
F	F	T	T	T	T	T	T
F	F	F	T	T	T	T	T

6. Show that $P \vee (Q \wedge R)$ and $(P \vee Q) \wedge (P \vee R)$ are logically equivalent.

Solution.
From columns (6) and (8), we have $P \vee (Q \wedge R)$ and $(P \vee Q) \wedge (P \vee R)$ are logically equivalent.

Truth Table showing the given Formulas are Logically Equivalent

P	Q	R	$P \vee Q$	$P \vee R$	$(P \vee Q) \wedge (P \vee R)$	$Q \wedge R$	$P \vee (Q \wedge R)$
T	T	T	T	T	T	T	T
T	T	F	T	T	T	F	T
T	F	T	T	T	T	F	T
F	T	T	T	T	T	T	T
T	F	F	T	T	T	F	T
F	T	F	T	F	F	F	F
F	F	T	F	T	F	F	F
F	F	F	F	F	F	F	F

7. Show that the propositions $P \rightarrow Q$ and $\neg P \vee Q$ are logically equivalent.

Solution.
From columns (3) and (5) in the table below, we have $P \rightarrow Q$ and $\neg P \vee Q$ are logically equivalent.

Truth Table showing the given
Formulas are Logically Equivalent

P	Q	$P \rightarrow Q$	$\neg P$	$\neg P \vee Q$
T	T	T	F	T
T	F	F	F	F
F	T	T	T	T
F	F	T	T	T

1.7 Inference Theory

Given a set of premises $H_1, H_2, \ldots, H_m$ and a conclusion C, we want to show whether C logically follows from $H_1, H_2, \ldots, H_m$.

That is, we want to show $(H_1 \wedge H_2 \wedge \cdots \wedge H_m) \rightarrow C$ is a tautology.

Procedure 1.
We look for those rows of $H_1, H_2, \ldots, H_m$ which have a truth value T. If for every such row C also has a truth value T, then the conclusion C logically follows from $H_1, H_2, \ldots, H_m$.

Example 1: $H_1 : P, \quad H_2 : P \rightarrow Q, \quad C : Q.$
In the first row of the truth table,

Truth Table Showing $P, P \to Q \Rightarrow Q$

P	**Q**	**P → Q**
T	T	T
T	F	F
F	T	T
F	F	T

$H_1 : P$ is True, $H_2 : P \to Q$ is True. Also, $C : Q$ is True.
Therefore, $P, P \to Q \Rightarrow Q$.

1.7.1 Rules of Inference

1. **Rule P**: A premise can be introduced at any point of derivation.

2. **Rule T**: A formula can be introduced provided it is tautologically implied by previously introduced formulas in the derivation.

3. **Rule CP**: If S can be derived from R and a set of premises, then $R \to S$ can be derived from the set of premises alone.

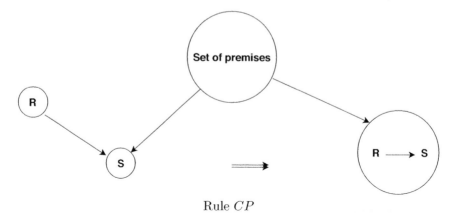

Rule CP

We use the following tables of implications and equivalences.
Implications Table

$$I_1 \; : \; P \Rightarrow P \vee Q$$
$$I_2 \; : \; Q \Rightarrow P \vee Q$$
$$I_3 \; : \; P \wedge Q \Rightarrow P$$
$$I_4 \; : \; P \wedge Q \Rightarrow Q$$
$$I_5 \; : \; P, P \to Q \Rightarrow Q$$
$$I_6 \; : \; \neg Q, P \to Q \Rightarrow \neg P$$

I_7 : $\neg P, P \vee Q \Rightarrow Q$

I_8 : $P \to Q, Q \to R \Rightarrow P \to R$

I_9 : $P, Q \Rightarrow P \vee Q$

I_{10} : $Q \Rightarrow P \to Q$

I_{11} : $P \vee Q, Q \to R \Rightarrow \neg P \to R$

I_{12} : $\neg P \Rightarrow P \to Q$.

Equivalences Table

E_1 : $\neg\neg P \Leftrightarrow P$

E_2 : $P \to Q \Leftrightarrow \neg P \vee Q$

E_3 : $P \to Q \Leftrightarrow \neg Q \to \neg P$

E_4 : $(P \rightleftharpoons Q) \Leftrightarrow (P \to Q) \wedge (Q \to P)$

E_5 : $P \to (Q \to R) \Leftrightarrow (P \wedge Q) \to R$

E_6 : $\neg(P \wedge Q) \Leftrightarrow \neg P \vee \neg Q$.

Note:

1. Rule CP means rule of Conditional Proof.

2. Rule CP is also called the deduction theorem.

3. In general, whenever conclusion is of the form $R \to S$ (in terms of conditional), we should apply Rule CP. In such case, R is taken as an additional premise, and S can be derived from the given premises and R.

1.7.2 Solved Problems

1. Show that $R \wedge (P \vee Q)$ is a valid conclusion from the premises $P \vee Q$, $Q \to R$, $P \to M$, and $\neg M$.

    Solution.
    Given premises are $P \vee Q$, $Q \to R$, $P \to M$, $\neg M$
    Conclusion: $R \wedge (P \vee Q)$.

$\{1\}$	(1)	$P \to M$	Rule P
$\{2\}$	(2)	$\neg M$	Rule P
$\{1, 2\}$	(3)	$\neg P$	Rule T $[\neg Q, P \to Q \Rightarrow \neg P]$
$\{4\}$	(4)	$P \vee Q$	Rule P
$\{4\}$	(5)	$\neg P \to Q$	Rule T $[P \to Q \Leftrightarrow \neg P \vee Q]$
$\{1, 2, 4\}$	(6)	Q	Rule T $[P, P \to Q \Rightarrow Q]$
$\{7\}$	(7)	$Q \to R$	Rule P
$\{1, 2, 4, 7\}$	(8)	R	Rule T $[P, P \to Q \Rightarrow Q]$
$\{1, 2, 4, 7\}$	(9)	$R \wedge (P \vee Q)$	Rule T $[P, Q \Rightarrow P \wedge Q]$.

2. Prove that the premises $P \rightarrow Q$, $Q \rightarrow R$, $R \rightarrow S$, $S \rightarrow \neg R$, and $P \wedge S$ are inconsistent.

Solution.

{1}	(1)	$P \rightarrow Q$	Rule P
{2}	(2)	$Q \rightarrow R$	Rule P
{1,2}	(3)	$P \rightarrow R$	Rule T $[P \rightarrow Q, Q \rightarrow R \Rightarrow P \rightarrow R]$
{4}	(4)	$S \rightarrow \neg R$	Rule P
{4}	(5)	$R \rightarrow \neg S$	Rule T $[P \rightarrow Q \Leftrightarrow \neg Q \rightarrow \neg P]$
{1,2,4}	(6)	$P \rightarrow \neg S$	Rule T $[P \rightarrow R, R \rightarrow \neg S \Rightarrow P \rightarrow \neg S]$
{1,2,4}	(7)	$\neg P \vee \neg S$	Rule T $[P \rightarrow Q \Leftrightarrow \neg P \vee Q]$
{1,2,4}	(8)	$\neg(P \wedge S)$	Rule T [De Morgan's law]
{9}	(9)	$P \wedge S$	Rule P
{1,2,4,9}	(10)	$(P \wedge S) \wedge$ $\neg(P \wedge S)$	Rule T $[P, Q \Rightarrow P \wedge Q]$

which is false. Therefore, the given set of premises are inconsistent.

3. Show that $(p \rightarrow q) \wedge (r \rightarrow s)$, $(q \rightarrow t) \wedge (s \rightarrow u)$, $\neg(t \wedge u)$, and $(p \rightarrow r) \Rightarrow \neg p$.

Solution.

{1}	(1)	$(p \rightarrow q) \wedge (r \rightarrow s)$	Rule P
{1}	(2)	$p \rightarrow q$	Rule T $[P \wedge Q \Rightarrow P]$
{1}	(3)	$r \rightarrow s$	Rule T $[P \wedge Q \Rightarrow Q]$
{4}	(4)	$(q \rightarrow t) \wedge (s \rightarrow u)$	Rule P
{4}	(5)	$q \rightarrow t$	Rule T $[P \wedge Q \Rightarrow P]$
{4}	(6)	$s \rightarrow u$	Rule T $[P \wedge Q \Rightarrow Q]$
{1,4}	(7)	$p \rightarrow t$	Rule T $[P \rightarrow Q,$ $Q \rightarrow R \Rightarrow P \rightarrow R]$
{1,4}	(8)	$r \rightarrow u$	Rule T $[P \rightarrow Q,$ $Q \rightarrow R \Rightarrow P \rightarrow R]$
{9}	(9)	$p \rightarrow r$	Rule P
{1,4,9}	(10)	$p \rightarrow u$	Rule T $[P \rightarrow Q,$ $Q \rightarrow R \Rightarrow P \rightarrow R]$
{1,4,9}	(11)	$\neg u \rightarrow \neg p$	Rule T
{1,4}	(12)	$\neg t \rightarrow \neg p$	Rule T
{1,4,9}	(13)	$(\neg t \vee \neg u) \rightarrow \neg p$	Rule T $[P \rightarrow Q,$ $R \rightarrow Q \Rightarrow (P \vee R) \rightarrow Q]$
{1,4,9}	(14)	$\neg(t \wedge u) \rightarrow \neg p$	Rule T [De Morgan's law]

$\{15\}$	(15)	$\neg(t \wedge u)$	Rule P
$\{1, 4, 9, 15\}$	(16)	$\neg p$	Rule T $[P, P \to Q \Rightarrow Q]$.

4. Show that the hypotheses, "It is not sunny this afternoon and it is colder than yesterday", "We will go swimming only if it is sunny", "If we do not go swimming, then we will take a canoe trip", and "if we take a canoe trip, then we will be home by sunset", lead to the conclusion, "We will be home by sunset".

Solution.

Let A: It is sunny
   B: It is colder than yesterday
   C: We will go swimming
   D: We will take a canoe trip
   E: We will be home by sunset.

Then the given premises are
(1) $\neg A \wedge B$ (2) $A \to C$ (3) $\neg C \to D$ (4) $D \to E$.
The conclusion is C.

$\{1\}$	(1)	$\neg A \wedge B$	Rule P
$\{1\}$	(2)	$\neg A$	Rule T $[P \wedge Q \Rightarrow P]$
$\{2\}$	(3)	$A \to C$	Rule P
$\{1, 2\}$	(4)	$\neg C$	Rule T $[\neg P, P \to Q \Rightarrow \neg Q]$
$\{5\}$	(5)	$\neg C \to D$	Rule P
$\{1, 2, 5\}$	(6)	D	Rule T $[P, P \to Q \Rightarrow Q]$
$\{7\}$	(7)	$D \to E$	Rule P
$\{1, 2, 5, 7\}$	(8)	E	Rule T $[P, P \to Q \Rightarrow Q]$.

5. Prove that the following argument is valid.

$$p \to \neg q, r \to q, r \Rightarrow \neg p.$$

Solution.

$\{1\}$	(1)	r	Rule P
$\{2\}$	(2)	$r \to q$	Rule P
$\{1, 2\}$	(3)	q	Rule T $[P, P \to Q \Rightarrow Q]$
$\{4\}$	(4)	$p \to \neg q$	Rule P
$\{1, 2, 4\}$	(5)	$\neg p$	Rule T $[P \to \neg Q, q \Rightarrow \neg P]$.

6. Using indirect method of proof, derive $p \to \neg s$ from the premises $p \to (q \vee r)$, $q \to \neg p$, $s \to \neg r$, and p.

Solution.

Let us include $\neg(p \to \neg s)$ as an additional premise and prove this problem by the method of contradiction.

Now, $\neg(p \to \neg s) = \neg(\neg p \lor \neg s) = p \land s$.

Therefore, the additional premise is $p \land s$.

$\{1\}$	(1)	$p \land s$	Additional premise
$\{2\}$	(2)	$p \to (q \lor r)$	Rule P
$\{3\}$	(3)	p	Rule P
$\{2,3\}$	(4)	$q \lor r$	Rule T $[P, P \to Q \Rightarrow Q]$
$\{1,2,3\}$	(5)	s	Rule T $[P \land Q \Rightarrow Q]$
$\{6\}$	(6)	$s \to \neg r$	Rule P
$\{1,2,3,6\}$	(7)	$\neg r$	Rule T $[P, P \to Q \Rightarrow Q]$
$\{1,2,3,6\}$	(8)	q	Rule T $[P, P \to Q \Rightarrow Q]$
$\{9\}$	(9)	$q \to \neg p$	Rule P
$\{1,2,3,6,9\}$	(10)	$\neg p$	Rule P
$\{1,2,3,6,9\}$	(11)	$p \land \neg p$	Rule T $[P, Q \Rightarrow P \land Q]$

which is false. Therefore, by the method of contradiction, $p \to \neg s$ follows.

7. Show that $R \to S$ can be derived from the premises $P \to (Q \to S)$, $\neg R \lor P$, and Q.

Solution.

$\{1\}$	(1)	R	Assumed premise
$\{2\}$	(2)	$\neg R \lor P$	Rule P
$\{2\}$	(3)	$R \to P$	Rule T $[P \to Q \Leftrightarrow \neg P \lor Q]$
$\{1,2\}$	(4)	P	Rule T $[P, P \to Q \Rightarrow Q]$
$\{5\}$	(5)	$P \to (Q \to S)$	Rule P
$\{1,2,5\}$	(6)	$Q \to S$	Rule P $[P, P \to Q \Rightarrow Q]$
$\{7\}$	(7)	Q	Rule P
$\{1,2,5,7\}$	(8)	S	Rule T $[P, P \to Q \Rightarrow Q]$
$\{1,2,5,7\}$	(9)	$R \to S$	Rule CP.

8. Prove that $A \to \neg D$ is a conclusion from the premises $A \to B \lor C$, $B \to \neg A$, and $D \to \neg C$ by using conditional proof.

Solution.

$\{1\}$	(1)	A	Assumed premise
$\{2\}$	(2)	$A \to B \lor C$	Rule P

$\{1,2\}$	(3)	$B \vee C$	Rule T $[P, P \rightarrow Q \Rightarrow Q]$
$\{1,2\}$	(4)	$\neg B \rightarrow C$	Rule T
$\{5\}$	(5)	$B \rightarrow \neg A$	Rule P
$\{5\}$	(6)	$A \rightarrow \neg B$	Rule T $[P \rightarrow Q \Leftrightarrow \neg Q \rightarrow \neg P]$
$\{1,2,5\}$	(7)	$A \rightarrow C$	Rule T
			$[P \rightarrow Q, Q \rightarrow R \Rightarrow P \rightarrow R]$
$\{8\}$	(8)	$D \rightarrow \neg C$	Rule P
$\{8\}$	(9)	$C \rightarrow \neg D$	Rule T $[P \rightarrow Q \Leftrightarrow \neg P]$
$\{1,2,5,8\}$	(10)	$A \rightarrow \neg D$	Rule CP.

1.8 Indirect Method of Proof

1.8.1 Method of Contradiction

In order to show that a conclusion C follows logically from the premises $H_1, H_2, \ldots, H_m$, we assume that C is false and consider $\neg C$ as an additional premise. If the new set of premises gives contradicting value, then the assumption $\neg C$ is true does not hold simultaneously with $H_1 \wedge H_2 \wedge \cdots \wedge H_m$ being true.

Therefore, C is true whenever $H_1 \wedge H_2 \wedge \cdots \wedge H_m$ is true. Thus, C logically follows from the premises $H_1 \wedge H_2 \wedge \cdots \wedge H_m$.

1.8.2 Solved Problems

1. Show that $\sqrt{2}$ is irrational.

   Solution.
   Suppose $\sqrt{2}$ is irrational.

   Therefore, $2 = \frac{p}{q}$ for $p, q \in \mathbb{Z}$, $q \neq 0$, p and q have no common divisor.

   Therefore, $\frac{p^2}{q^2} = 2 \implies p^2 = 2q^2$.

   Since p^2 is an even integer, p is an even integer.

   Let $p = 2m$ for some integer m.

   $$\therefore \quad p^2 = 2q^2 \implies (2m)^2 = 2q^2$$
   $$\implies q^2 = 2m^2.$$

   Since q^2 is an even integer, q is an even integer.

   Let $q = 2n$ for some integer n.

Hence, p and q are even integers.

Hence, they can have a common factor 2 which is a contradiction to our assumption.

$\therefore$ $\sqrt{2}$ is irrational.

2. Prove that the following argument is valid:

    "If 7 is less than 4, then 7 is not a prime number; 7 is not less than 4; therefore 7 is a prime number".

    Solution.

    Let A: 7 is less than 4.

        B: 7 is a prime number.

    The given premises are   (i) $A \rightarrow \neg B$ (ii) $\neg A$, and the conclusion is B.

$\{1\}$	(1)	$\neg A$	Rule P
$\{2\}$	(2)	$A \rightarrow \neg B$	Rule P
$\{1,2\}$	(3)	$\neg(\neg B)$	Rule T $[\neg P, P \rightarrow Q \Rightarrow \neg Q]$
$\{1,2\}$	(4)	B	

    Therefore, the statements are valid.

3. Prove that the sum of an irrational number and a rational number is irrational.

    Solution.

    Let a be a rational number and b be an irrational number.

    Assume $a + b = c$ is a rational number.

    $\therefore$ $b = c + (-a)$ is a rational number since the sum of two rational numbers is a rational number, which is a contradiction to our assumption.

    Hence, the sum of an irrational number and a rational number is an irrational number.

4. Show that the statement "Every positive integer is the sum of the squares of three integers" is false.

    Solution.

    Any number of the form $4m + 7$, where m is non-negative integer, cannot be written as the sum of the squares of three integers.

    For example, $7, 11, 15, 19, \ldots$ cannot be written as the sum of squares of three integers.

    $\therefore$ "Every positive integer is the sum of the squares of three integers" is a false statement.

1.9 Method of Contrapositive

In order to prove that $H_1 \wedge H_2 \wedge \cdots \wedge H_m \Rightarrow C$, if we prove $\neg C \Rightarrow \neg(H_1 \wedge H_2 \wedge \cdots \wedge H_m)$, then the original problem follows. This procedure is called the method of contrapositive.

1.9.1 Solved Problems

1. Prove that if $3n + 2$ is odd, then n is odd.

   Solution.
   Let us prove this problem by the method of contrapositive.
   Assume that n is even. Then $n = 2k$, for some integer k.
   Now,
   $$3n + 2 = 3(2k) + 2 = 2(3k + 1)$$

   $\implies$ $3n + 2$ is an even number which contradicts that $3n + 2$ is odd.
   $\therefore$ n is odd.

2. Prove that if n is an integer and $n^3 + 5$ is odd, then n is even.

   Solution.
   Given: $n^3 + 5$ is odd.
   To prove: n is even.

   Assume that n is odd.
   $\therefore$ $n = 2k + 1$ for some integer k.
   $$\begin{aligned} n^3 + 5 &= (2k + 1)^3 + 5 \\ &= 8k^3 + 12k^2 + 6k + 6 \\ &= 2\left(4k^3 + 6k^2 + 3k + 3\right) \end{aligned}$$

   $\implies$ $n^3 + 5$ is an even number which is a contradiction.
   $\therefore$ n is even.

1.10 Various Methods of Proof

1.10.1 Trivial Proof

In an implication $p \to q$, if we can establish that q is true, then regardless of the truth value of p, the implication $p \to q$ will be true.

Hence, to construct a trivial proof of $p \to q$, we need to show that the truth value of q is true.

1.10.2   Vacuous Proof

If the hypothesis p of an implication $p \to q$ is false, then $p \to q$ is true for any proposition q.

Example:
Prove the proposition $p(0)$ where $p(n)$ is the proposition: "If n is a positive integer greater than 1, then $n^2 > n$".

Solution.
Let $p(0)$: If 0 is a positive integer greater than 1, then $0^2 > 0$.
Since 0 is not a positive integer greater than 1, the proposition is true.

1.10.3   Direct Proof

Suppose, the hypothesis p is true. Then, the implication $p \to q$ can be proved if we can prove that q is true by using the rules of inference and some other theorems.

Example:
Prove that the sum of two odd integers is even.

Solution.
Let a and b be two odd integers.
$\therefore$ $a = 2m + 1$, for some integer m.
$b = 2n + 1$, for some integer n.
$\therefore$ $a + b = 2m + 2n + 2 = 2(m + n + 1)$ $\implies$ $a + b$ is even.
$\therefore$ Sum of two odd integers is even.

1.11   Predicate Calculus

Consider the following statement:
Amruta is a student. Suppose "Amruta" is taken as a and "is a student" as S, then the above statement can be symbolically written as $S(a)$. Here, "is a student" is a predicate, and "Amruta" is a subject.

Any statement "r is P" can be written as $P(r)$. This can be extended to two place predicate. For example, Amruta is taller than Arvindh. Similarly, a statement with three place predicate can be extended as "Amruta is younger than Boomika but elder than Arvindh".

If a statement function of one variable is defined to be an expression consisting of a predicate symbol and an individual variable, such a statement function becomes a statement when the variable is replaced by the name of the object. Example for statement function of two variables:

$A(x, y)$: x is taller than y.
$G(a, r)$: Amruta is taller than Arvindh where a is Amruta and r is Arvindh.

1.11.1   Quantifiers

Consider the following example:

All apples are red. This can be understood as "for any x, if x is an apple, then x is red".

If we denote $A(x)$: x is an apple and $R(x)$: x is red,

then we can write the above statement as

$$(x)(A(x) \to R(x)).$$

Here (x) is called "Universal Quantifier". We use universal quantifier for those statements of the form "All P are Q".

Now, consider the following example:

Some men are clever. This can be written as "There exists x; if x is a man, then he is clever".

If we denote $M(x)$: x is a man and $C(x)$: x is clever,

then we can write the above statement as

$$(\exists x)(M(x) \wedge C(x)).$$

Here $(\exists x)$ is called "Existential Quantifier". Existential quantifier is used for those statements which are of the form "Some P are Q".

1.11.2   Universe of Discourse, Free and Bound Variables

Consider the following statement:

All men are giants. This can be symbolically written as

$$(x)(M(x) \to G(x))$$

where $M(x)$: x is a man and $G(x)$: x is giant.

In the above example, if we restrict the class as the class of men, then the symbolic representation will be $(x)G(x)$. Such a restricted class is called "Universe of Discourse".

In any formula, the part containing $(x)P(x)$ or $\exists x P(x)$ is called the x bound part of the formula. Any variable appearing in an x bound part of the formula is called bound variable. Otherwise, it is called free. Any formula immediately following (x) or $(\exists x)$ is called the scope of the quantifier.

Example:

$$(x)P(x) \wedge Q(x)$$

In this, all x in $P(x)$ is bound, whereas the x in $Q(x)$ is free. The scope of (x) is $P(x)$.

1.11.3   Solved Problems

1. Symbolise the expression "x is the father of the mother of y".

   Solution.

   Let $P(x)$: x is a person

       $F(x, y)$: x is the father of y

       $M(x, y)$: x is the mother of y.

   Let z be the mother of y. Hence, the given statement can be written as

   "x is the father of z, and z is the mother of y".

   In symbolic notation,

   $$(\exists z)(P(z) \wedge F(x, z) \wedge M(z, y)).$$

2. Given $P = \{2, 3, 4, 5, 6\}$, state the truth value of the statement $(\exists x \in P)(x + 3 = 10)$.

   Solution.

   For $x = 2, 3, 4, 5, 6$, no x satisfies $x + 3 = 10$.

       $\therefore$ $(\exists x \in P)(x + 3 = 10)$ is false.

3. Write the symbolic form, and negate the following statements:

   (i) Everyone who is healthy can do all kinds of work.

   (ii) Some people are not admired by everyone.

   (iii) Everyone should help his neighbours, or his neighbours will not help him.

   Solution.

   (i) Let $H(x)$: x is healthy

         $W(x)$: x can do all kinds of work.

       Symbolic form of the statement is

   $$(x)(H(x) \rightarrow W(x)).$$

   Negation of this expression is

   $$\neg((x)(H(x) \rightarrow W(x))$$
   $$\Longrightarrow \neg((x)(\neg H(x) \vee W(x)))$$
   $$\Longrightarrow (\exists x)(H(x) \wedge \neg W(x)).$$

   That is, someone is healthy and cannot do all kinds of work.

(ii) Let $A(x)$: x is admired.

Then, the given statement can be written as "for some x, it is not a case that x is admired by everyone".

Symbolic form is

$$(\exists x)(\neg A(x)).$$

Negation of the above statement is

$$\neg((\exists x)\neg A(x)) \Rightarrow (\forall x)A(x).$$

That is, all people are admired by everyone.

(iii) This statement can be rewritten as

"For all x, x is a person, x should help his neighbours, or his neighbours will not help him".

Let $P(x)$: x is a person

$H(x)$: x helps his neighbour.

The symbolic form is

$$(x)((P(x) \to H(x)) \vee (H(x) \to \neg P(x))).$$

Negation of the above statement is

$$(\exists x)((H(x) \to P(x)) \wedge (\neg P(x) \to H(x))).$$

4. Find a counterexample, if possible to these universally quantified statements, in which the universe of discourse for all variables consists of all integers.

(i) $\forall x \, \forall y \, (x^2 = y^2 \to x = y).$

(ii) $\forall x \, \forall y \, (xy \geq x).$

Solution.

(i) Suppose $x = -3$ and $y = 3$, then $x^2 = y^2 = 9$.

But $x \neq y$.

$\therefore \, \forall x \, \forall y \, (x^2 = y^2 \to x = y)$ is false.

(ii) Suppose $x = -3$ and $y = 3$, then $xy = -9$.

$-9 < -3 \implies xy < x.$

$\therefore \, \forall x \, \forall y (xy > x)$ is false.

5. Establish this logical equivalence, where A is a proposition not involving any quantifiers. Show that $(\forall x p(x)) \wedge A \equiv \forall x(p(x) \wedge A)$ and $(\exists x \, p(x)) \wedge A \equiv \exists x(p(x) \wedge A).$

Solution.

(a) Consider

$$(\forall x p(x)) \wedge A \equiv \forall x(p(x) \wedge A). \tag{1.3}$$

Case (i): Suppose A is true.

   Since $P \wedge T = P$, equation (1.3) $\implies (\forall x p(x)) \wedge T = (\forall x) p(x)$.
   Therefore, both sides are the same, and hence equation (1.3)
   is logically equivalent.

Case (ii): Suppose A is false.

   Since $P \wedge F = F$, the left-hand side of equation (1.3) is false.
   Also, for every x, $p(x) \wedge A$ is false. Hence, right-hand side of
   equation (1.3) is false. Therefore, both sides are the same,
   and hence, it is logically equivalent.

   (b) Now, consider

$$(\exists x p(x) \wedge A) = \exists x (p(x) \wedge A). \tag{1.4}$$

Case (i): Suppose A is true.

   Since $P \wedge T = P$, equation (1.4) becomes $\exists x p(x) = (\exists x) p(x)$,
   and hence, equation (1.4) is logically equivalent.

Case (ii): Suppose A is false.

   Since $P \wedge F = F$, the left-hand side of equation (1.4) is false.
   Also, for every x, $p(x) \wedge A$ is false. Therefore, $\exists x (p(x) \wedge A)$
   is false. Hence, the two sides of equation (1.4) are the same,
   and hence equation (1.4) is logically equivalent.

6. Show that $\exists x P(x) \wedge \exists x Q(x)$ and $\exists x (P(x) \wedge Q(x))$ are not logically
    equivalent.

Solution.

We prove the two statements are not logically equivalent by using
the following counterexample.

Let us assume that the universe of discourse is the set of integers.

Let $P(x)$: x is a positive integer

   $Q(x)$: x is a negative integer.

Then, $\exists x P(x) \wedge \exists x Q(x)$ is true.

But the truth value of $\exists x (P(x) \wedge Q(x))$ is false.

$\therefore$ Given two statements are not logically equivalent.

7. Use quantifiers and predicate to express the fact that $\lim\limits_{x \to a} f(x)$ does
    not exist.

Solution.

Limit exists means "$f(x)$ approaches L (where $L \in \mathbb{R}$) as
x approaches a, if given $\epsilon > 0$, $\exists \ \delta > 0$ such that
$|f(x) - L| < \epsilon$ whenever $0 < |x - a| < \delta$".

Limit does not exist means "for all real numbers L, $\lim\limits_{x \to a} f(x) \neq L$".

The above statement can be expressed as

$$\neg(\forall\, \epsilon > 0, \exists\, \delta > 0,\ \forall x(0 < |x - a| < \delta) \to |f(x) - L| < \epsilon))$$
$$\implies \forall\, L, \exists \epsilon > 0, \forall\, \delta > 0, (\exists\, x)() < |x - a| < \delta \wedge |f(x) - L| < \epsilon)$$
$$[\because \neg(p \to q) = p \wedge \neg q].$$

8. Let $H = \{-1, 0, 1, 2\}$ denote the universe of discourse. If $p(x, y) : x + y = 1$, find the truth value of $(\exists\, x)(\exists\, y)\ p(x, y)$.

   Solution.

   When $x = -1, \exists y = 2$ such that $-1 + 2 = 1$.

   When $x = 0, \exists y = 1$ such that $0 + 1 = 1$.

   When $x = 1, \exists y = 0$ such that $1 + 0 = 1$.

   When $x = 2, \exists y = -1$ such that $2 - 1 = 1$.

   $\therefore$ $(\forall x)(\exists y)\ p(x, y)$ is true.

9. What are the negations of the statements $\forall x(x^2 > x)$ and $\exists x(x^2 = 2)$?

   Solution.
   (i) Given statement is $\forall x(x^2 > x)$.
       Its negation is $\exists x(x^2 \leq x)$.
   (ii) Given statement is $\exists x(x^2 = 2)$.
       Its negation is $\forall x(x^2 \neq 2)$.

10. Write the negation of the statement $\exists x\ (\forall y)\ p(x, y)$.

    Solution.

    Given statement is $(\exists x)\ (\forall y)\ p(x, y)$.

    Its negation is $\neg((\exists x)\ (\forall y)\ p(x, y)) \implies (\forall x)\ (\exists y)\neg p(x, y)$.

11. Consider the statement "Given any positive integer, there is a greater positive integer". Symbolize this statement using and without using the set of positive integers as the universe of discourse.

    Solution.
    Let $G(x, y) : x$ is greater than y.

    If we use the universe of discourse as the set of positive integers, then we can write $(x)\ (\exists y)\ G(y, x)$.

    If we do not impose the restriction on the universe of discourse and if we write $P(x) : x$ is a positive integer, then we can write as

    $$(x)\ (P(x) \to (\exists y)\ (P(y) \wedge G(y, x))).$$

12. Indicate free and bound variables. Also indicate the scope of the quantifier in

   (i) $(x)\ (P(x) \wedge R(x)) \Rightarrow (x)\ (P(x)) \wedge Q(x)$.
   (ii) $((x)(P(x) \rightleftharpoons Q(x) \wedge \exists(x)\ R(x))) \wedge S(x)$.

Solution.

(i) All occurrences of x in $P(x) \wedge R(x)$ are bound occurrences. The occurrence of x in $xP(x)$ is bound. The occurrence of x in $Q(x)$ is free. The scope of (x) is $P(x) \wedge R(x)$ and $P(x)$.

(ii) All occurrences of x in $P(x) \rightleftharpoons Q(x) \wedge (\exists\ x)\ R(x)$ are bound, and the occurrence of x in $S(x)$ is free. The scope of (x) is $P(x) \rightleftharpoons Q(x) \wedge (\exists x)\ R(x)$, and the scope of $(\exists x)$ is $R(x)$.

1.11.4   Inference Theory for Predicate Calculus

We have seen implication table and equivalence table already. Those rules can be extended here also. For example, $P, P \rightarrow Q \Rightarrow Q$ can be extended as $P(x), P(x) \rightarrow Q(x) \Rightarrow Q(x)$.
In addition, we use the following rules:

1. Universal Specification (US): $(x)A(x) \Rightarrow A(y)$
2. Universal Generalization (UG): $A(y) \Rightarrow (x)A(x)$
3. Existential Specification (ES): $(\exists x)A(x) \Rightarrow A(y)$
4. Existential Generalization (EG): $A(y) \Rightarrow (\exists x)A(x)$.

Remark.
Let the universe of discourse be denoted by $S = \{a_1, a_2, \ldots, a_n\}$. Then,

$$(x)A(x) = A(a_1) \wedge A(a_2) \wedge \cdots \wedge A(a_n)$$
$$(\exists x)A(x) = A(a_1) \vee A(a_2) \vee \cdots \vee A(a_n).$$

Consider now
$$\neg(x)A(x)$$
$$\Leftrightarrow \neg(A(a_1) \wedge A(a_2) \wedge \cdots \wedge A(a_n))$$
$$\Leftrightarrow \neg A(a_1) \vee \neg A(a_2) \vee \cdots \vee \neg A(a_n)$$
$$\Leftrightarrow (\exists x)\neg A(x).$$

Similarly, $\neg((\exists x)A(x)) \Leftrightarrow (x)\neg A(x))$.

1.11.5   Solved Problems

1. Show that
   $$(x)(P(x) \rightarrow Q(x)) \wedge (x)(Q(x) \rightarrow R(x)) \Rightarrow (x)(P(x) \rightarrow R(x)).$$

Solution.

$\{1\}$	(1)	$(x)(P(x) \rightarrow Q(x))$	Rule P
$\{2\}$	(2)	$P(y) \rightarrow Q(y)$	Rule US
$\{3\}$	(3)	$(x)(Q(x) \rightarrow R(x))$	Rule P
$\{3\}$	(4)	$Q(y) \rightarrow R(y)$	Rule US
$\{1,3\}$	(5)	$P(y) \rightarrow R(y)$	Rule T
			$[\because P \rightarrow Q, Q \rightarrow R \Rightarrow P \rightarrow R]$
$\{1,3\}$	(6)	$(x)(P(x) \rightarrow R(x))$	Rule UG.

2. Show that $\forall x(P(x) \vee Q(x)) \Rightarrow \forall x\ P(x) \vee \exists x\ Q(x)$ using indirect method.

Solution.

We use the method of contradiction. Assume $\neg((x)P(x) \vee (\exists x)Q(x))$ as an additional premise.

$\{1\}$	(1)	$\neg((x)P(x) \vee (\exists x)Q(x))$	Assumed premise
$\{1\}$	(2)	$(\exists x)\neg P(x) \wedge (x)\neg Q(x)$	Rule T [De Morgan's law]
$\{1\}$	(3)	$(\exists x)\neg P(x)$	Rule T $[P \wedge Q \Rightarrow P]$
$\{1\}$	(4)	$(x)\neg Q(x)$	Rule T $[P \wedge Q \Rightarrow Q]$
$\{1\}$	(5)	$\neg P(y)$	Rule ES
$\{1\}$	(6)	$\neg Q(y)$	Rule US
$\{1\}$	(7)	$\neg P(y) \wedge \neg Q(y)$	Rule T $[P, Q \Rightarrow P \wedge Q]$
$\{1\}$	(8)	$\neg(P(y) \vee Q(y))$	Rule T [De Morgan's law]
$\{9\}$	(9)	$(x)(P(x) \vee Q(x))$	Rule P
$\{9\}$	(10)	$P(y) \vee Q(y)$	Rule US
$\{1,9\}$	(11)	$(P(y) \vee Q(y))\wedge$ $\neg(P(y) \vee Q(y))$	Rule T $[P, Q \Rightarrow P \wedge Q]$

which is false.

Therefore, by the method of contradiction, we have

$$\forall x(P(x) \vee Q(x)) \Rightarrow \forall x\ P(x) \vee \exists x\ Q(x).$$

3. Show that $\forall x P(x) \wedge \exists x Q(x)$ is equivalent to $\forall x \exists y(P(x) \wedge Q(y))$.

Solution.

$\{1\}$	(1)	$\forall x P(x) \wedge \exists x Q(x)$	Rule P
$\{1\}$	(2)	$\forall x P(x)$	Rule T $[P \wedge Q \Rightarrow P]$
$\{1\}$	(3)	$P(m)$	Rule US
$\{1\}$	(4)	$\exists x Q(x)$	Rule T $[P \wedge Q \Rightarrow Q]$

$\{1\}$ (5) $Q(n)$ Rule ES

$\{1\}$ (6) $P(m) \wedge Q(n)$ Rule T

$\{1\}$ (7) $\exists y(P(m) \wedge Q(y))$ Rule EG

$\{1\}$ (8) $(\forall x)\exists y(P(x) \wedge Q(y))$ Rule UG

Hence, $\forall x P(x) \wedge \exists x Q(x)$ and $\forall x \exists y(P(x) \wedge Q(y))$ are logically equivalent.

4. Show that $(x)(P(x) \to Q(x)) \Rightarrow (x)P(x) \to (x)Q(x)$.

   Solution.

   We use contrapositive method to prove this problem.

$\{1\}$ (1) $\neg((x)P(x) \to (x)Q(x))$ Assumed premise

$\{1\}$ (2) $(x)P(x) \wedge \neg((x)Q(x))$ Rule T $[\neg(P \to Q) \Leftrightarrow P \wedge \neg Q]$

$\{1\}$ (3) $(x)P(x)$ Rule T $[P \wedge Q \Rightarrow P]$

$\{1\}$ (4) $\neg((x)Q(x))$ Rule T $[P \wedge Q \Rightarrow Q]$

$\{1\}$ (5) $(\exists x)\neg Q(x)$ Rule T [apply $\neg$]

$\{1\}$ (6) $P(y)$ Rule US

$\{1\}$ (7) $\neg Q(y)$ Rule ES

$\{1\}$ (8) $P(y) \wedge \neg Q(y)$ Rule T $[P, Q \Rightarrow P \wedge Q]$

$\{1\}$ (9) $\neg(P(y) \to Q(y))$ Rule T $[P \wedge \neg Q \Leftrightarrow \neg(P \to Q)]$

$\{1\}$ (10) $(\exists x)\neg(P(x) \to Q(x))$ Rule EG

$\{1\}$ (11) $\neg((x)(P(x) \to Q(x)))$ Rule T [apply $\neg$].

$\therefore$ By the method of contrapositive, we have

$$(x)(P(x) \to Q(x)) \Rightarrow (x)P(x) \to (x)Q(x).$$

5. Show that $\exists x(P(x) \wedge Q(x)) \Rightarrow (\exists x)P(x) \wedge (\exists x)Q(x)$. Is the converse true?

   Solution.

$\{1\}$ (1) $(\exists x)(P(x) \wedge Q(x))$ Rule P

$\{1\}$ (2) $P(y) \wedge Q(y)$ Rule ES

$\{1\}$ (3) $P(y)$ Rule T $[P \wedge Q \Rightarrow P]$

$\{1\}$ (4) $Q(y)$ Rule T $[P \wedge Q \Rightarrow Q]$

$\{1\}$ (5) $\exists x P(x)$ Rule EG

$\{1\}$ (6) $\exists x Q(x)$ Rule EG

$\{1\}$ (7) $\exists x P(x) \wedge \exists x Q(x)$ Rule T $[P, Q \Rightarrow P \wedge Q]$.

Converse is also true, since

{1}	(1)	$(\exists x)\,P(x) \wedge (\exists x)Q(x)$	Rule P
{1}	(2)	$(\exists x)\,P(x)$	Rule T $[P \wedge Q \Rightarrow P]$
{1}	(3)	$(\exists x)Q(x)$	Rule T $[P \wedge Q \Rightarrow Q]$
{1}	(4)	$P(y)$	Rule ES
{1}	(5)	$Q(y)$	Rule ES
{1}	(6)	$P(y) \wedge Q(y)$	Rule T $[P, Q \Rightarrow P \wedge Q]$
{1}	(7)	$(\exists x)(P(x) \wedge Q(x))$	Rule EG.

6. Verify the validity of the following argument. Every living thing is a plant or an animal. John's gold fish is alive, and it is not a plant. All animals have hearts. Therefore, John's gold fish has a heart.

Solution.

Let $L(x)$: x is a living thing

$P(x)$: x is a plant

$A(x)$: x is an animal

$H(x)$: x has a heart.

Given premises with their symbolic forms are

(i) Every living thing is a plant or an animal.
$$(x)(L(x) \rightarrow P(x) \vee A(x)).$$

(ii) John's gold fish is alive, and it is not a plant.
$$L(y) \wedge \neg P(y).$$

(iii) All animals have hearts.
$$(x)(A(x) \rightarrow H(x)).$$

Conclusion is $H(y)$.

{1}	(1)	$(x)(L(x) \rightarrow P(x) \vee A(x))$	Rule P
{1}	(2)	$L(y) \rightarrow P(y) \vee A(y)$	Rule P
{3}	(3)	$L(y) \wedge \neg P(y)$	Rule P
{4}	(4)	$L(y)$	Rule T $[P \wedge Q \Rightarrow P]$
{5}	(5)	$\neg P(y)$	Rule T $[P \wedge Q \Rightarrow Q]$
{1,3}	(6)	$P(y) \vee A(y)$	Rule T $[P, P \rightarrow Q \Rightarrow Q]$

$\{1,3\}$	(7)	$\neg P(y) \rightarrow A(y)$	Rule T $[P \rightarrow Q \Rightarrow \neg P \vee Q]$
$\{8\}$	(8)	$(x)(A(x) \rightarrow H(x))$	Rule P
$\{8\}$	(9)	$A(y) \rightarrow H(y)$	Rule US
$\{1,3,8\}$	(10)	$\neg P(y) \rightarrow H(y)$	Rule T
			$[P \rightarrow Q, Q \rightarrow R \Rightarrow P \rightarrow R]$
$\{1,3,8\}$	(11)	$H(y)$	Rule T $[P \rightarrow Q, P \Rightarrow Q]$.

Hence, the given statements are valid statements.

1.12   Additional Solved Problems

1. Let $P(x)$ denote the statement $x \leq 4$. Write the truth values of $P(2)$ and $P(4)$.

   Solution.

   $$P(x) : x \leq 4$$
   $$P(2) : 2 \leq 4 \quad \text{is True}$$
   $$P(6) : 6 \leq 4 \quad \text{is False.}$$

2. Give the converse and the contrapositive of the implication: "If it is raining, then I get wet".

   Solution.

   Let P: It is raining

   $\quad\quad Q$: I get wet.

   Given statement is $P \rightarrow Q$.

   Its inverse is

   $Q \rightarrow P$: If I get wet, then it is raining.

   Its contrapositive is

   $\neg Q \rightarrow \neg P$: If I do not get wet, then it is not raining.

3. Show that $R \vee S$ follows logically from the premises $C \vee D$, $C \vee D \rightarrow \neg H$, $\neg H \rightarrow (A \wedge \neg B)$, and $(A \wedge \neg B) \rightarrow (R \vee S)$.

   Solution.

$\{1\}$	(1)	$C \vee D \rightarrow \neg H$	Rule P
$\{2\}$	(2)	$\neg H \rightarrow (A \wedge \neg B)$	Rule P
$\{1,2\}$	(3)	$C \vee D \rightarrow A \wedge \neg B$	Rule T
$\{4\}$	(4)	$(A \wedge \neg B) \rightarrow (R \vee S)$	Rule P
$\{1,2,4\}$	(5)	$C \vee D \rightarrow R \vee S$	Rule T

$$\{6\} \quad (6) \qquad\qquad C \vee D \qquad \text{Rule } P$$
$$\{1,2,4,6\} \quad (7) \qquad\qquad R \vee S \qquad \text{Rule } T \; [P \rightarrow Q, P \Rightarrow Q].$$

4. Show that the following premises are inconsistent.

   (i) If Jack misses many classes through illness, then he fails high school.
   (ii) If Jack fails high school, then he is uneducated.
   (iii) If Jack reads a lot of books, then he is not uneducated.
   (iv) Jack misses many classes through illness and reads a lot of books.

   Solution.

   $$E : \text{Jack misses many classes}$$
   $$S : \text{Jack fails high school}$$
   $$A : \text{Jack reads a lot of books}$$
   $$H : \text{Jack is uneducated}$$

   The given premises are

   $$E \rightarrow S, \; S \rightarrow H, \; A \rightarrow \neg H, \text{ and } \; E \wedge A.$$

$\{1\}$	(1)	$E \rightarrow S$	Rule P
$\{2\}$	(2)	$S \rightarrow H$	Rule P
$\{1,2\}$	(3)	$E \rightarrow H$	Rule T
$\{4\}$	(4)	$A \rightarrow \neg H$	Rule P
$\{4\}$	(5)	$H \rightarrow \neg A$	Rule T
$\{1,2,4\}$	(6)	$E \rightarrow \neg A$	Rule T
$\{1,2,4\}$	(7)	$\neg E \wedge \neg A$	Rule T
$\{1,2,4\}$	(8)	$\neg(E \wedge A)$	Rule T
$\{9\}$	(9)	$E \wedge A$	Rule P
$\{1,2,4,9\}$	(10)	$(E \wedge A) \wedge (\neg(E \wedge A))$	Rule T

   which is false. Hence, the given premises are inconsistent.

5. Write the dual of $(P \wedge Q) \vee T$.

   Solution.

   $(P \vee Q) \wedge F$.

6. Negate the following statements.

   (i) Ottawa is a small town.
   (ii) Every city in Canada is clean.

Solution.

(i) Ottawa is not a small town.

(ii) Every city in Canada is not clean.

7. Construct the truth table for $P \wedge (P \vee Q)$.

Solution.

The truth table is shown below.

Truth Table for $P \wedge (P \vee Q)$

P	Q	$P \vee Q$	$P \wedge (P \vee Q)$
T	T	T	T
T	F	T	T
F	T	T	F
F	F	F	F

8. Write the following in symbolic form:

If John takes calculus or Peter takes analytical geometry, then Mohan will take English.

Solution.

Let J: John takes calculus

$\quad P$: Peter takes analytical geometry

$\quad M$: Mohan will take English.

Then the symbolic form is $(J \vee P) \to M$.

9. Symbolise the expression: "All the world loves a lover".

Solution.

Let $P(x)$: x is a person

$\quad L(x)$: x is a lover

$\quad R(x, y)$: x loves y.

The symbolic form is $(x)(P(x) \to (y)(P(y) \wedge L(y) \to R(x, y))$.

10. Obtain the PDNF and PCNF of

$$P \to ((P \to Q) \wedge \neg(\neg Q \vee \neg P)).$$

Solution.

$$\text{Let} \quad A \Leftrightarrow P \to ((P \to Q) \wedge \neg(\neg Q \vee \neg P))$$

$$\Leftrightarrow \neg P \vee ((\neg P \vee Q) \wedge (Q \wedge P))$$

$$\Leftrightarrow \neg P \vee ((\neg P \wedge (Q \wedge P)) \vee (Q \wedge (Q \wedge P)))$$

$$\Leftrightarrow \neg P \vee F \vee (P \wedge Q)$$

$$\Leftrightarrow \neg P \vee (P \wedge Q)$$

$$\Leftrightarrow \vee (P \wedge Q)$$
$$\Leftrightarrow (\neg P \wedge (Q \vee \neg Q)) \vee (P \wedge Q)$$
$$\Leftrightarrow (\neg P \wedge Q) \vee (\neg P \wedge \neg Q) \vee (P \wedge Q)$$

which is the required PDNF. Its PCNF is

$$\neg \neg A \Leftrightarrow \neg P \vee Q.$$

11. Obtain the PCNF of $S : (\neg P \to R) \wedge (Q \rightleftharpoons P)$. Hence, obtain PDNF.

Solution.

$$S : (\neg P \to R) \wedge (Q \rightleftharpoons P)$$
$$\Leftrightarrow (\neg P \to R) \wedge ((Q \to P) \wedge (P \to Q))$$
$$\Leftrightarrow (P \vee R) \wedge (\neg Q \vee P) \wedge (\neg P \vee Q)$$
$$\Leftrightarrow ((P \vee R) \vee F) \wedge ((\neg Q \vee P) \vee F) \wedge ((\neg P \vee Q) \vee F)$$
$$\Leftrightarrow ((P \vee R) \vee (Q \wedge \neg Q)) \wedge ((\neg Q \vee P) \vee (R \wedge \neg R))$$
$$\wedge ((\neg P \vee Q) \vee (R \wedge \neg R))$$
$$\Leftrightarrow (P \vee Q \vee R) \wedge (P \vee \neg Q \vee R) \wedge (P \vee \neg Q \vee R)$$
$$\wedge (P \vee \neg Q \vee \neg R) \wedge (\neg P \vee Q \vee R) \wedge (\neg P \vee Q \vee \neg R)$$
$$\Leftrightarrow (P \vee Q \vee R) \wedge (P \vee \neg Q \vee R) \wedge (P \vee \neg Q \vee \neg R)$$
$$\wedge (\neg P \vee Q \vee R) \wedge (\neg P \vee Q \vee \neg R)$$

which is the required PCNF. Now,

$$\neg S : (P \vee Q \neg R) \wedge (\neg P \vee \neg Q \vee R) \wedge (\neg P \vee \neg Q \neg R)$$
$$\neg \neg S : \neg ((P \vee Q \neg R) \wedge (\neg P \vee \neg Q \vee R) \wedge (\neg P \vee \neg Q \neg R))$$
$$\Leftrightarrow (\neg P \wedge \neg Q \wedge R) \vee (P \wedge Q \wedge R) \vee (P \wedge Q \wedge R)$$

which is the required PDNF.

12. What is the contrapositive statement of "The home town wins whenever it is raining"?

Solution.

Let P: It is raining

Q: The home town wins.

Given statement is $P \to Q$.

Its contrapositive statement is

$\neg Q \to \neg P$: If the home town does not win, then it is not raining.

13. Give the symbolic form of "Some men are giants".

     Solution.

     The given statement can be written as

     "there is an x such that x is a man and x is giant".

     Let $M(x)$: x is a man

     $\quad G(x)$: x is a giant.

     $\quad \therefore$ The symbolic form is

     $$(\exists x)(M(x) \wedge G(x)).$$

14. Find the PCNF of $(P \vee R) \wedge (P \vee \neg Q)$. Also find its PDNF, without using truth table.

     Solution.

     $$\begin{aligned}
     \text{Let} \quad A &\Leftrightarrow (P \vee R) \wedge (P \vee \neg Q)\\
     &\Leftrightarrow ((P \vee R) \vee F) \wedge ((P \vee \neg Q) \vee F)\\
     &\Leftrightarrow ((P \vee R) \vee (Q \wedge \neg Q)) \wedge ((P \vee \neg Q) \vee (R \wedge \neg R))\\
     &\Leftrightarrow (P \vee Q \vee R) \wedge (P \vee \neg Q \vee R) \wedge (P \vee \neg Q \vee R)\\
     &\qquad \wedge (P \vee \neg Q \wedge \neg R)\\
     &\Leftrightarrow (P \vee Q \vee R) \wedge (P \vee \neg Q \vee R) \wedge (P \vee \neg Q \vee \neg R)
     \end{aligned}$$

     which is the required PCNF. Now,

     $$\begin{aligned}
     \neg A : &(P \vee Q \vee \neg R) \wedge (\neg P \vee Q \vee R) \wedge (\neg P \vee Q \vee \neg R)\\
     &\wedge (\neg P \vee \neg Q \vee R) \wedge (\neg P \vee \neg Q \vee \neg R)\\
     \neg \neg A : &\neg\big((P \vee Q \vee \neg R) \wedge (\neg P \vee Q \vee R) \wedge (\neg P \vee Q \vee \neg R)\\
     &\wedge (\neg P \vee \neg Q \vee R) \wedge (\neg P \vee \neg Q \vee \neg R)\big)\\
     &\Leftrightarrow (\neg P \wedge \neg Q \wedge R) \vee (P \wedge \neg Q \wedge \neg R) \vee (P \wedge \neg Q \wedge R)\\
     &\quad \vee (P \wedge Q \wedge \neg R) \vee (P \wedge Q \wedge R)
     \end{aligned}$$

     which is the required PDNF.

15. Show that using rule CP, $\neg P \vee Q$, $\neg Q \vee R$, $R \to S \Leftrightarrow P \to S$.

     Solution.

$\{1\}$	(1)	P	Assumed premise
$\{2\}$	(2)	$\neg P \vee Q$	Rule P
$\{2\}$	(3)	$P \to Q$	Rule T $[P \to Q \Leftrightarrow \neg P \vee Q]$
$\{1,2\}$	(4)	Q	Rule T $[P, P \to Q \Rightarrow Q]$
$\{5\}$	(5)	$\neg Q \vee R$	Rule P
$\{5\}$	(6)	$Q \to R$	Rule T $[P \to Q \Leftrightarrow \neg P \vee Q]$

$\{1,2,5\}$	(7)	R	Rule T $[P, P \rightarrow Q \Rightarrow Q]$
$\{8\}$	(8)	$R \rightarrow S$	Rule P
$\{1,2,5,8\}$	(9)	S	Rule T $[P, P \rightarrow Q \Rightarrow Q]$
$\{1,2,5,8\}$	(10)	$P \rightarrow S$	Rule CP.

16. Show that $(\neg P \wedge (\neg Q \wedge R)) \vee (Q \wedge R) \vee (P \wedge R) \Leftrightarrow R$ without using truth table.

Solution.

$$\neg P \wedge (\neg Q \wedge R) \Leftrightarrow (\neg P \wedge \neg Q) \wedge R \quad \text{(Associative law)}$$
$$\Leftrightarrow \neg(P \vee Q) \wedge R \quad \text{(De Morgan's law).} \quad (1.5)$$

Now,

$$(Q \wedge R) \vee (P \wedge R) \Leftrightarrow (Q \vee P) \wedge R \quad \text{(Distributive law)}$$
$$\Leftrightarrow (P \vee Q) \wedge R \quad \text{(Commutative law).} \quad (1.6)$$

From (1.5) and (1.6), we have

$$(\neg P \wedge (\neg Q \wedge R)) \vee (Q \wedge R) \vee (P \wedge R)$$
$$\Leftrightarrow (\neg(P \vee Q) \wedge R) \vee ((P \vee Q) \wedge R)$$
$$\Leftrightarrow (\neg(P \vee Q) \vee (P \vee Q)) \wedge R \quad \text{(Distributive law)}$$
$$\Leftrightarrow T \wedge R$$
$$\Leftrightarrow R.$$

$$\therefore \ (\neg P \wedge (\neg Q \wedge R)) \vee (Q \wedge R) \vee (P \wedge R) \Leftrightarrow R.$$

2

Combinatorics

2.1   Introduction

In this chapter, we discuss about the technique of mathematical induction which is used for proving many standard results over natural numbers. Then, we discuss about the basis of counting, pigeonhole principle, permutations and combinations, and recurrence relation. These concepts are useful in the analysis of certain discrete time systems, analysis of algorithms, error-correcting code, etc. At the end of the chapter, we discuss about generating function which is used to solve linear recurrence relations.

2.2   Mathematical Induction

Mathematical induction is a method of finding the truth from a general statement for particular cases.

A statement may be true with reference to more than hundred cases yet we cannot conclude it to be true in general. It is possible to disprove the statement by a counter-example. A statement need not be accepted to be true. Such a statement inferred from a particular case is called a conjecture. If the conjecture is a statement involving natural numbers, we can use the principle of the mathematical induction to prove the same.

2.2.1   Principle of Mathematical Induction

For a given statement involving a natural number n, if we can show that

(1)   The statement is true for $n = 1$ or $n = n_0$,

(2)   The statement is true for $n = m + 1$ under the assumption that the statement is true for $n = m (m \geq n_0)$,

then we can conclude that the statement is true for all natural numbers.

Remark 2.2.1 *In the above principle, (1) is usually referred to as the basis of induction, and (2) is usually referred to as the induction step. Also, the assumption that the statement is true for $n = m$ in (2) is usually referred to as the induction hypothesis.*

2.2.2   Procedure to Prove that a Statement $P(n)$ is True for all Natural Numbers

Step 1.   We must prove that $P(1)$ is true.

Step 2.   By assuming $P(m)$ is true, we must prove that $P(m+1)$ is also true.

In the sequel, we apply the principle of mathematical induction to prove statements involving natural numbers.

2.2.3   Solved Problems

1. Prove that $1 + 2 + 3 + \cdots + n = \dfrac{n(n+1)}{2}$ by induction principle.

   Solution.

   Let $P(n) : 1 + 2 + 3 + \cdots + n = \dfrac{n(n+1)}{2}$.

   (1) $P(1) : 1 = \dfrac{1(1+1)}{2} = 1$ is true.

   (2) Assume that $P(m)$ is true.

   That is, $1 + 2 + 3 + \cdots + m = \dfrac{m(m+1)}{2}$.

   (3) Now,

   $$
   \begin{aligned}
   1 + 2 + 3 + \cdots + m + (m+1) &= \frac{m(m+1)}{2} + m + 1 \\
   &= \frac{m(m+1) + 2(m+1)}{2} \\
   &= \frac{(m+1)(m+2)}{2}.
   \end{aligned}
   $$

   $\therefore$ By mathematical induction, the given statement is true for all n.

2. Show that $1^2 + 2^2 + 3^2 + \cdots + n^2 = \dfrac{n(n+1)(2n+1)}{6}$, $n \geq 1$ by mathematical induction.

   Solution.

   Let $P(n) : 1^2 + 2^2 + 3^2 + \cdots + n^2 = \dfrac{n(n+1)(2n+1)}{6}$.

   (1) $P(1) : 1^2 = \dfrac{1(1+1)(2+1)}{6} = 1$ is true.

   (2) Assume $P(m)$ is true.

   That is, $1^2 + 2^2 + 3^2 + \cdots + m^2 = \dfrac{m(m+1)(2m+1)}{6}$.

   (3) Now,

   $$
   1^2 + 2^2 + 3^2 + \cdots + m^2 + (m+1)^2
   $$

$$= \frac{m(m+1)(2m+1)}{6} + (m+1)^2$$

$$= \frac{m(m+1)(2m+1) + 6(m+1)^2}{6}$$

$$= \frac{(m+1)[m(2m+1) + 6(m+1)]}{6}$$

$$= \frac{(m+1)(2m^2 + m + 6m + 6)}{6}$$

$$= \frac{(m+1)(2m^2 + 7m + 6)}{6}$$

$$= \frac{(m+1)(2m^2 + 4m + 3m + 6)}{6}$$

$$= \frac{(m+1)[2m(m+2) + 3(m+2)]}{6}$$

$$= \frac{(m+1)(m+2)(m+3)}{6}$$

$$= \frac{(m+1)[(m+1) + 1][2(m+1) + 1]}{6}.$$

∴ By mathematical induction, the given statement is true for all $n \geq 1$.

3. Prove that $1^3 + 2^3 + 3^3 + \cdots + n^3 = \dfrac{n^2(n+1)^2}{4}$, $n \in \mathbb{N}$.

Solution.

Let $P(n) : 1^3 + 2^3 + 3^3 + \cdots + n^3 = \dfrac{n^2(n+1)^2}{4}$.

(1) $P(1) : 1^2 = \dfrac{1^2(1+1)^2}{4}$ is true.

(2) Assume $P(m)$ is true.

That is, $1^3 + 2^3 + 3^3 + \cdots + m^3 = \dfrac{m^2(m+1)^2}{4}$.

(3) Now,

$$1^3 + 2^3 + 3^3 + \cdots + m^3 + (m+1)^3 = \frac{m^2(m+1)^2}{4} + (m+1)^3$$

$$= \frac{m^2(m+1)^2 + 4(m+1)^3}{4}$$

$$= \frac{(m+1)^2[m^2 + 4(m+1)]}{4}$$

$$= \frac{(m+1)^2(m^2 + 4m + 4)}{4}$$

$$= \frac{(m+1)^2(m+2)^2}{4}$$

$$= \frac{(m+1)[(m+1) + 1]^2}{4}.$$

$\therefore$ By mathematical induction, the statement is true for all $n \in \mathbb{N}$.

4. Prove that $\sum_{i=1}^{n}(2i-1)^2 = n^2$, for all $n \in \mathbb{N}$.

(or)

Prove that the sum of the first n odd integers is n^2 for all integers n.

Solution.

Let $P(n) : 1 + 3 + 5 + \cdots + (2n-1) = n^2$.

(1) $P(1) : 1 = 1^2$ is true.

(2) Assume $P(m)$ is true.

That is, $1 + 3 + 5 + \cdots + (2m-1) = m^2$.

(3) Now,

$$
\begin{aligned}
1 + 3 + 5 + \cdots + (2m-1) + [2(m+1)-1] &= m^2 + [2(m+1)-1] \\
&= m^2 + [2(m+1)-1] \\
&= m^2 + 2m + 2 - 1 \\
&= m^2 + 2m + 1 \\
&= (m+1)^2.
\end{aligned}
$$

$\therefore$ By mathematical induction, the given statement is true for all $n \in \mathbb{N}$.

(or)

(1) $P(1) : 1 = 1^2$ is true.

(2) Assume $P(m)$ is true.

That is, $\sum_{i=1}^{m}(2i-1) = m^2$.

(3) Now,

$$
\begin{aligned}
\sum_{i=1}^{m+1}(2i-1) &= \sum_{i=1}^{m}(2i-1) + [2(m+1)-1] \\
&= m^2 + 2m + 2 - 1 \\
&= m^2 + 2m + 1 \\
&= (m+1)^2.
\end{aligned}
$$

$\therefore$ By mathematical induction, the given statement is true for all $n \in \mathbb{N}$.

5. Using mathematical induction, prove that

$$
2 + 5 + 8 + \cdots + (3n-1) = \frac{n(3n+1)}{2}.
$$

Solution.

Let $P(n) : 2 + 5 + 8 + \cdots + (3n-1) = \dfrac{n(3n+1)}{2}$.

(1) $P(1) : 2 = \dfrac{1[3 \cdot 1 + 1]}{2} = 2$ is true.

(2) Assume $P(m)$ is true.

That is, $2 + 5 + 8 + \cdots + (3m - 1) = \dfrac{m(3m + 1)}{2}$.

(3) Now,

$$2 + 5 + 8 + \cdots + (3m - 1) + [3(m + 1) - 1]$$
$$= \frac{m(3m + 1)}{2} + (3m + 3 - 1)$$
$$= \frac{m(3m + 1) + 2(3m + 2)}{2}$$
$$= \frac{3m^2 + m + 6m + 4}{2}$$
$$= \frac{3m^2 + 7m + 4}{2}$$
$$= \frac{3m^2 + 3m + 4m + 4}{2}$$
$$= \frac{3m(m + 1) + 4(m + 1)}{2}$$
$$= \frac{(m + 1)(3m + 4)}{2}$$
$$= \frac{(m + 1)[3(m + 1) + 1]}{2}.$$

∴ By mathematical induction, the given statement is true.

6. Prove that for $n \geq 0$, $1 + 2 + 4 + \cdots + 2^n = 2^{n+1} - 1$.

Solution.

Let $P(n) : 1 + 2 + 4 + \cdots + 2^n = 2^{n+1} - 1$.

(1) $P(1) : 1 + 2 = 2^{1+1} - 1$ is true.

(2) Assume $P(m)$ is true.

That is, $1 + 2 + 4 + \cdots + 2^m = 2^{m+1} - 1$.

(3) Now,

$$1 + 2 + 4 + \cdots + 2^m + 2^{m+1} = (2^{m+1} - 1) + 2^{m+1}$$
$$= 2 \cdot 2^{m+1} - 1$$
$$= 2^{m+2} - 1$$
$$= 2^{(m+1)+1} - 1.$$

∴ By mathematical induction, the given statement is true.

7. Prove that if $n \geq 1$, then

$$1(1!) + 2(2!) + \cdots + n(n!) = (n + 1)! - 1.$$

Solution.

Let $P(n) : 1(1!) + 2(2!) + \cdots + n(n!) = (n+1)! - 1$.

(1) $P(1) : 1(1!) = (1+1)! - 1$ is true.

(2) Assume $P(m)$ is true.

That is, $1(1!) + 2(2!) + \cdots + m(m!) = (m+1)! - 1$.

(3) Now,

$$1(1!) + 2(2!) + \cdots + m(m!) + (m+1)[(m+1)!]$$
$$= [(m+1)! - 1] + (m+1)[(m+1)!]$$
$$= (m+1)! + (m+1)(m+1)! - 1$$
$$= (m+1)![1 + (m+1)] - 1$$
$$= (m+1)!(m+2) - 1$$
$$= (m+2)! - 1.$$

∴ By mathematical induction, the given statement is true for all $n \geq 1$.

8. Use mathematical induction to show that

$$\frac{1}{1 \cdot 2} + \frac{1}{2 \cdot 3} + \frac{1}{3 \cdot 4} + \cdots + \frac{1}{n(n+1)} = \frac{n}{n+1}, \quad \text{for all } n \geq 1.$$

Solution.

Let $P(n) : \dfrac{1}{1 \cdot 2} + \dfrac{1}{2 \cdot 3} + \dfrac{1}{3 \cdot 4} + \cdots + \dfrac{1}{n(n+1)} = \dfrac{n}{n+1}$.

(1) $P(1) : \dfrac{1}{1 \cdot 2} = \dfrac{1}{1+1}$ is true.

(2) Assume $P(m)$ is true.

That is, $\dfrac{1}{1 \cdot 2} + \dfrac{1}{2 \cdot 3} + \dfrac{1}{3 \cdot 4} + \cdots + \dfrac{1}{m(m+1)} = \dfrac{m}{m+1}$.

(3) Now,

$$\frac{1}{1 \cdot 2} + \frac{1}{2 \cdot 3} + \frac{1}{3 \cdot 4} + \cdots + \frac{1}{m(m+1)} + \frac{1}{(m+1)(m+2)}$$
$$= \frac{m}{m+1} + \frac{1}{(m+1)(m+2)}$$
$$= \frac{m(m+2) + 1}{(m+1)(m+2)}$$
$$= \frac{m^2 + 2m + 1}{(m+1)(m+2)}$$
$$= \frac{(m+1)^2}{(m+1)(m+2)}$$
$$= \frac{m+1}{(m+1)+1}.$$

∴ By mathematical induction, the given statement is true for all n.

9. Use mathematical induction to prove that $n^3 - n$ is divisible by 3 whenever n is a positive integer.

Solution.

Let $P(n) : n^3 - n$ is divisible by 3.

(1) $P(1) : 1^1 - 1 = 0$ is divisible by 3.
(2) Assume $P(m)$ is true.
   That is, $m^3 - m$ is divisible by 3.
(3) Now,

$$(m+1)^3 - (m+1) = (m^3 + 3m^2 + 3m + 1) - (m+1)$$
$$= (m^3 - m) + 3m^2 + 3m$$
$$= (m^3 - m) + 3(m^2 + m).$$

Since both terms in this sum are divisible by 3, it follows that $(m+1)^3 - (m+1)$ is also divisible by 3.

$\therefore$ By mathematical induction, $n^3 - n$ is divisible by 3 for all $n \in \mathbb{N}$.

10. Use mathematical induction to show that $n^3 + 2n$ is divisible by 3 whenever n is a non-negative integer.

(or)

Prove that for all $n \geq 1$, $n^3 + 2n$ is a multiple of 3.

Solution.

Let $P(n) : n^3 + 2n$ is a multiple of 3.

(1) $P(1) : 1^3 + 2 \cdot 1 = 3$ is a multiple of 3.
(2) Assume $P(m)$ is true.
   That is, $m^3 + 2m$ is a multiple of 3.
(3) Now,

$$(m+1)^3 + 2(m+1)$$
$$= m^3 + 3m^2 + 3m + 1 + 2m + 2$$
$$= (m^3 + 2m) + 3(m^2 + m + 1).$$

Since both terms on the right-hand side are divisible by 3, it follows that $(m+1)^3 + 2(m+1)$ is a multiple of 3.

Hence, by mathematical induction, $n^3 + 2n$ is a multiple of 3 for all $n \in \mathbb{N}$.

11. Use mathematical induction to show that $8^n - 3^n$ is a multiple of 5.

    Solution.

    Let $P(n) : 8^n - 3^n$ is a multiple of 5.

    (1) $P(1) : 8 - 3 = 5$ is a multiple of 5.
    (2) Assume $P(m)$ is true. That is, $8^m - 3^m$ is a multiple of 5.
    (3) Now,

    $$8^{m+1} - 3^{m+1} = 8^m(8) - 3^{m+1}$$
    $$= 8^m(5+3) - 3^{m+1}$$
    $$= 5 \cdot 8^m + 3 \cdot 8^m - 3^{m+1}$$
    $$= 5 \cdot 8^m + 3(8^m - 3^m).$$

    Since both terms on the right-hand side are multiples of 5, it follows that $8^{m+1} - 3^{m+1}$ is also a multiple of 5.

    $\therefore$ By mathematical induction, $8^n - 3^n$ is a multiple of 5.

12. Use mathematical induction to show that $n^2 - 7n + 12$ is non-negative whenever n is an integer greater than 3.

    Solution.

    Let $P(n) : n^2 - 7n + 12$ be non-negative.

    (1) $P(4) : 4^2 - 7(4) + 12 = 16 - 28 + 12 = 0$, which is non-negative.
    (2) Assume $P(m)$ is true.
        That is, $m^2 - 7m + 12$ is non-negative.
    (3) Now,

    $$(m+1)^2 - 7(m+1) + 12$$
    $$= m^2 + 2m + 1 - 7m - 7 + 12$$
    $$= (m^2 - 7m + 12) + (2m - 6).$$

    The first term is non-negative, and the second term is also non-negative for $m > 3$. Hence, $(m+1)^2 - 7(m+1) + 12$ is non-negative.

    $\therefore$ By mathematical induction, $n^2 - 7n + 12$ is non-negative for $n > 3$.

13. Use mathematical induction to prove the inequality $n < 2^n$ for all positive integers n.

    Solution.

    Let $P(n) : n < 2^n$.

    (1) $P(1) : 1 < 2^1$ is true.
    (2) Assume $P(m)$ is true.
        That is, $m < 2^m$.

(3) Now,

$$m + 1 < 2^m + 1$$
$$< 2^m + 2^m \quad \text{since } 1 < 2^m$$
$$< 2 \cdot 2^m$$
$$< 2^{m+1}.$$

∴ By mathematical induction, the result is true for all $n \in \mathbb{N}$.

14. Prove that $n + 10 \leq 2^n$ for all $n \in \mathbb{N}$ and $n \geq 4$.

Solution.

Let $P(n) : n + 10 \leq 2^n, n \geq 4$.

(1) $P(4) : 4 + 10 \leq 2^4$ is true.
(2) Assume $P(m)$ is true.
   That is, $m + 10 \leq 2^m$.
(3) Now,

$$(m + 1) + 10 = (m + 10) + 1$$
$$\leq 2^m + 1$$
$$\leq 2^m + 2^m \quad \text{since } 1 < 2^m$$
$$\leq 2 \cdot 2^m$$
$$\leq 2^{m+1}.$$

∴ By the principle of mathematical induction, the result is true for all $n \in \mathbb{N}$ and $n \geq 4$.

Note: In the above problem, the result is false for $n = 1, 2$ and 3.

15. Prove that $2^n < n!$ for $n \geq 4$ and $n \in \mathbb{N}$.

Solution.

Let $P(n) : 2^n < n!$ for $n \geq 4$.

(1) $P(4) : 2^4 < 4!$ is true (since $16 < 24$).
(2) Assume $P(m)$ is true.
   That is, $2^m < m!$
(3) Now,

$$2^{m+1} = 2^m \cdot 2$$
$$< 2^m(m + 1) \quad \text{since } 2 < m + 1 \quad \text{for } m \geq 4$$
$$< m!(m + 1) \quad \text{by assumption}$$
$$< (m + 1)!$$

∴ The result is true for all $n \geq 4$ by the principle of mathematical induction.

16. Suppose there are n people in a room ($n \geq 1$) and all shake hands with one another. Prove that $\dfrac{(n-1)n}{2}$ handshakes will have occurred.

    Solution.
    Let $P(n) : \dfrac{(n-1)n}{2}$ handshakes.

    (1) $P(1) : \dfrac{(1-1)1}{2} = 0$. There is no handshake when $n = 1$.
    (2) Assume $P(m)$ is true.
        That is, there are $\dfrac{(m-1)m}{2}$ handshakes.
    (3) Now, if one more person enters the room, he will shake hands with m people. So,

    $$\frac{(m-1)m}{2} + m = \frac{m^2 - m + 2m}{2}$$
    $$= \frac{m^2 + m}{2}$$
    $$= \frac{m(m+1)}{2}$$
    $$= \frac{[(m+1)-1](m+1)}{2}.$$

    $\therefore$ By the principle of mathematical induction, the result follows.

17. Use mathematical induction to prove that $3^n + 7^n - 2$ is divisible by 8, for $n \geq 1$.

    Solution.
    Let $P(n) : 3^n + 7^n - 2$ is divisible by 8.

    (1) $P(1) : 3^1 + 7^1 - 2 = 8$ is divisible by 8, which is true.
    (2) Assume $P(m)$ is true.
        That is, $3^m + 7^m - 2$ is divisible by 8.
    (3) Now,

    $$3^{m+1} + 7^{m+1} - 2 = 3 \cdot 3^m + 7 \cdot 7^m - 2$$
    $$= 3 \cdot 3^m + (3+4) \cdot 7^m - 2$$
    $$= 3 \cdot 3^m + 3 \cdot 7^m + 4 \cdot 7^m - 6 + 4$$
    $$= 3(3^m + 7^m - 2) + 4(7^m + 1).$$

    Since both terms on right-hand side are divisible by 8, $3^{m+1} + 7^{m+1} - 2$ is also divisible by 8.

    $\therefore$ By the principle of mathematical induction, the result follows.

18. Show that $a^n - b^n$ is divisible by $a - b$.

   Solution.
   Let $P(n) : a^n - b^n$ be divisible by $a - b$.

   (1) $P(1) : a^1 - b^1$ is divisible by $a - b$.
   (2) Assume $P(m)$ is true.
      That is, $a^m - b^m$ is divisible by $a - b$.

   $$\Rightarrow \quad a^m - b^m = k(a - b)$$

   $$\Rightarrow \quad a^m = b^m + k(a - b). \tag{2.1}$$

   (3) Now,

   $$\begin{aligned}
   a^{m+1} - b^{m+1} &= a^m \cdot a - b^m \cdot b \quad \text{[using (2.1)]} \\
   &= ak(a - b) + ab^m - b \cdot b^m \\
   &= (a - b)ak + (a - b)b^m \\
   &= (a - b)[ak + b^m], \quad \text{which is a multiple of } (a - b).
   \end{aligned}$$

   Hence, $a^{m+1} - b^{m+1}$ is divisible by $a - b$.

   $\therefore$ By the principle of mathematical induction, $a^n - b^n$ is divisible by $a - b$ for all $n \geq 1$.

19. Using mathematical induction, prove that
   $$2 + 2^2 + 2^3 + \cdots + 2^n = 2^{n+1} - 2.$$

   Solution.
   Let $P(n) : 2 + 2^2 + 2^3 + \cdots + 2^n = 2^{n+1} - 2$.

   (1) $P(1) : 2^1 = 2^{1+1} - 2 = 2$ is true.
   (2) Assume $P(m)$ is true.
      That is, $2 + 2^2 + 2^3 + \cdots + 2^m = 2^{m+1} - 2$ is true.
   (3) Now,

   $$\begin{aligned}
   2 + 2^2 + 2^3 + \cdots + 2^m + 2^{m+1} &= 2^{m+1} - 2 + 2^{m+1} \\
   &= 2 \cdot 2^{m+1} - 2 \\
   &= 2^{m+2} - 2.
   \end{aligned}$$

   $\therefore$ By the principle of mathematical induction, the result is true for all n.

20. Use mathematical induction to show that $n! \geq 2^{n+1}$; $n = 5, 6, \ldots$

    Solution.

    Let $P(n) : n! \geq 2^{n+1}$; $n = 5, 6 \ldots$

    (1) $P(5) : 5! \geq 2^{5+1}$ is true since $120 \geq 64$.

    (2) Assume $P(m)$ is true.

    That is, $\qquad\qquad m! \geq 2^{m+1}$; $m = 5, 6 \ldots \qquad\qquad$ (2.2)

    (3) Multiplying both sides of (2.2) by 2, we have

    $$2(m!) \geq 2 \cdot 2^{m+1}$$
    $$\implies (m+1)m! \geq 2^{m+2}$$
    $$\implies (m+1)! \geq 2^{m+2}.$$

    $\therefore$ By mathematical induction,

    $$n! \geq 2^{n+1}; \quad n = 5, 6, \ldots$$

21. Show that $\dfrac{1}{1 \cdot 2} + \dfrac{1}{2 \cdot 3} + \cdots + \dfrac{1}{n(n+1)} = \dfrac{n}{n+1}$.

    Solution.

    Let $P(n) : \dfrac{1}{1 \cdot 2} + \dfrac{1}{2 \cdot 3} + \cdots + \dfrac{1}{n(n+1)} = \dfrac{n}{n+1}$.

    (1) $P(1) : \dfrac{1}{1 \cdot 2} = \dfrac{1}{1+1}$ is true.

    (2) Assume $P(m)$ is true.

    That is, $\dfrac{1}{1 \cdot 2} + \dfrac{1}{2 \cdot 3} + \cdots + \dfrac{1}{m(m+1)} = \dfrac{m}{m+1}$.

    (3) Now,

    $$\frac{1}{1 \cdot 2} + \frac{1}{2 \cdot 3} + \cdots + \frac{1}{m(m+1)} + \frac{1}{(m+1)(m+2)}$$
    $$= \frac{m}{m+1} + \frac{1}{(m+1)(m+2)}$$
    $$= \frac{m(m+2) + 1}{(m+1)(m+2)}$$
    $$= \frac{m^2 + 2m + 1}{(m+1)(m+2)}$$
    $$= \frac{(m+1)^2}{(m+1)(m+2)}$$
    $$= \frac{m+1}{m+2}$$
    $$= \frac{m+1}{(m+1)+1}.$$

    $\therefore$ By the principle of mathematical induction, the result follows.

22. Using mathematical induction, prove that if $n \geq 1$, then

$$1 \cdot 1! + 2 \cdot 2! + 3 \cdot 3! + \cdots + n \cdot n! = (n+1)! - 1, \quad n \geq 1.$$

Solution.
Let $P(n) : 1 \cdot 1! + 2 \cdot 2! + 3 \cdot 3! + \cdots + n \cdot n! = (n+1)! - 1$.

(1) $P(1) : 1 \cdot 1! = (1+1)! - 1$ is true.
(2) Assume $P(m)$ is true.
  That is, $1 \cdot 1! + 2 \cdot 2! + 3 \cdot 3! + \cdots + m \cdot m! = (m+1)! - 1$.
(3) Now,

$$\begin{aligned}
1 \cdot 1! &+ 2 \cdot 2! + 3 \cdot 3! + \cdots + m \cdot m! + (m+1)(m+1)! \\
&= [(m+1)! - 1] + (m+1)(m+1)! \\
&= (m+1)![1 + m + 1] - 1 \\
&= (m+1)!(m+2) - 1 \\
&= (m+2)! - 1.
\end{aligned}$$

∴ By mathematical induction,

$$1 \cdot 1! + 2 \cdot 2! + 3 \cdot 3! + \cdots + n \cdot n! = (n+1)! - 1, \quad n \geq 1.$$

23. Using mathematical induction, prove that $\sum_{k=0}^{n} 3^k = \dfrac{3^{n+1} - 1}{2}$.

Solution.
Let $P(n) : 3^0 + 3^1 + \cdots + 3^n = \dfrac{3^{n+1} - 1}{2}$.

(1) $P(0) : 3^0 = \dfrac{3^1 - 1}{2} = \dfrac{2}{2}$ is true.
(2) Assume $P(m)$ is true.
  That is, $3^0 + 3^1 + \cdots + 3^m = \dfrac{3^{m+1} - 1}{2}$.
(3) Now,

$$\begin{aligned}
3^0 + 3^1 + \cdots + 3^m + 3^{m+1} &= \frac{3^{m+1} - 1}{2} + 3^{m+1} \\
&= \frac{3^{m+1} - 1 + 2 \cdot 3^{m+1}}{2} \\
&= \frac{3 \cdot 3^{m+1} - 1}{2} \\
&= \frac{3^{m+2} - 1}{2} \\
&= \frac{3^{(m+1)+1}}{2}.
\end{aligned}$$

∴ By mathematical induction, $\sum_{k=0}^{n} 3^k = \dfrac{3^{n+1} - 1}{2}$ is true.

24. Using mathematical induction, prove that

$$\frac{1}{\sqrt{1}} + \frac{1}{\sqrt{2}} + \frac{1}{\sqrt{3}} + \cdots + \frac{1}{\sqrt{n}} > \sqrt{n} \quad \text{for } n \geq 2.$$

Solution.

Let $P(n): \dfrac{1}{\sqrt{1}} + \dfrac{1}{\sqrt{2}} + \dfrac{1}{\sqrt{3}} + \cdots + \dfrac{1}{\sqrt{n}} > \sqrt{n} \quad \text{for } n \geq 2.$

(1) $P(2): \dfrac{1}{\sqrt{1}} + \dfrac{1}{\sqrt{2}} = 1.707 > \sqrt{2} = 1.414$ is true.

(2) Assume $P(m)$ is true.

That is, $\dfrac{1}{\sqrt{1}} + \dfrac{1}{\sqrt{2}} + \dfrac{1}{\sqrt{3}} + \cdots + \dfrac{1}{\sqrt{m}} > \sqrt{m} \quad \text{for } m \geq 2.$

(3) Now,

$$\frac{1}{\sqrt{1}} + \frac{1}{\sqrt{2}} + \frac{1}{\sqrt{3}} + \cdots + \frac{1}{\sqrt{m}} + \frac{1}{\sqrt{m+1}}$$

$$> \sqrt{m} + \frac{1}{\sqrt{m+1}}$$

$$> \frac{\sqrt{m}\sqrt{m+1} + 1}{\sqrt{m+1}}$$

$$> \frac{\sqrt{m(m+1)} + 1}{\sqrt{m+1}}$$

$$> \frac{\sqrt{m \cdot m} + 1}{\sqrt{m+1}} \quad (\because \ m+1 > m)$$

$$> \frac{\sqrt{m^2} + 1}{\sqrt{m+1}}$$

$$> \frac{m+1}{\sqrt{m+1}}$$

$$> \sqrt{m+1}.$$

$\therefore$ By the principle of mathematical induction,

$$\frac{1}{\sqrt{1}} + \frac{1}{\sqrt{2}} + \frac{1}{\sqrt{3}} + \cdots + \frac{1}{\sqrt{n}} > \sqrt{n} \quad \text{for } n \geq 2.$$

25. Using mathematical induction, prove that $n^3 + (n+1)^3 + (n+2)^3$ is divisible by 9, for $n \geq 1$.

Solution.

Let $P(n): n^3 + (n+1)^3 + (n+2)^3$ be divisible by 9, for $n \geq 1$.

(1) $P(1): 1^3 + (1+1)^3 + (1+2)^3 = 1 + 8 + 27 = 36$ is divisible by 9.

(2) Assume $P(m)$ is true.

That is, $m^3 + (m+1)^3 + (m+2)^3$ is divisible by 9, for $m \geq 1$.

(3) Now,

$$(m+1)^3 + [(m+1)+1]^3 + [(m+1)+2]^3$$
$$= (m+1)^3 + (m+2)^3 + (m+3)^3$$
$$= (m+1)^3 + (m+2)^3 + m^3 + 9m^2 + 27m + 27$$
$$[m^3 + (m+1)^3 + (m+2)^3] + 9(m^2 + 3m + 3).$$

Both the terms on the right-hand side are divisible by 9, and hence, the terms on the left-hand side are also divisible by 9.

∴ By the principle of mathematical induction, the result follows.

26. Show that $3^{2n} + 4^{n+1}$ is divisible by 5, for $n \geq 0$.

Solution.

Let $P(n) : 3^{2n} + 4^{n+1}$ be divisible by 5, for $n \geq 0$.

(1) $P(0) : 3^0 + 4^1 = 5$ is divisible by 5.

(2) Assume $P(m)$ is true.

That is, $3^{2m} + 4^{m+1}$ is divisible by 5.

$\Rightarrow \qquad 3^{2m} + 4^{m+1} = 5k$ (where k is an integer)

$\Rightarrow \qquad 3^{2m} = 5k - 4^{m+1}.$ $\hfill (2.3)$

(3) Now,

$$3^{2(m+1)} + 4^{m+1+1} = 3^{2m} \cdot 3^2 + 4^{m+2}$$
$$= (5k - 4^{m+1}) \cdot 3^2 + 4^{m+1} \cdot 4 \quad \text{[using (2.3)]}$$
$$= 5k \cdot 9 - 4^{m+1} \cdot 3^2 + 4^{m+1} \cdot 4$$
$$= 5k \cdot 9 - 4^{m+1} \cdot 9 + 4 \cdot 4^{m+1}$$
$$= 5k \cdot 9 - 5 \cdot 4^{m+1}$$
$$= 5(9k - 4^{m+1}) \quad \text{which is a multiple of 5.}$$

∴ By the principle of mathematical induction, $3^{2n} + 4^{n+1}$ is divisible by 5 for $n \geq 0$.

27. Using mathematical induction, prove that

$$H_{2^n} \geq 1 + \frac{n}{2} \quad \text{where} \quad H_k = 1 + \frac{1}{2} + \frac{1}{3} + \cdots + \frac{1}{k}.$$

Solution.

Let $P(n) : H_{2^n} = 1 + \dfrac{1}{2} + \dfrac{1}{3} + \cdots + \dfrac{1}{2^n} \geq 1 + \dfrac{n}{2}$.

(1) $P(0) : H_{2^0} = H_1 = 1 \geq 1 + \dfrac{0}{2}$.

That is, $1 \geq 1$ is true.

(2) Assume $P(m)$ is true.

That is, $P(m) : H_{2^m} = 1 + \dfrac{1}{2} + \dfrac{1}{3} + \cdots + \dfrac{1}{2^m} \geq 1 + \dfrac{m}{2}$.

(3) Now,

$$H_{2^{m+1}} = 1 + \frac{1}{2} + \frac{1}{3} + \cdots + \frac{1}{2^m} + \frac{1}{2^m + 1} + \frac{1}{2^m + 2} + \cdots + \frac{1}{2^{m+1}}$$

$$= H_{2^m} + \frac{1}{2^m + 1} + \frac{1}{2^m + 2} + \cdots + \frac{1}{2^{m+1}}$$

$$\geq \left(1 + \frac{m}{2}\right) + \frac{1}{2^m + 1} + \frac{1}{2^m + 2} + \cdots + \frac{1}{2^{m+1}}$$

$$\geq \left(1 + \frac{m}{2}\right) + 2^m \cdot \frac{1}{2^{m+1}} \quad \left(\text{since there are } 2^m \text{ terms}\right.$$

$$\left. \text{each not less than } \frac{1}{2^{m+1}}\right)$$

$$\geq \left(1 + \frac{m}{2}\right) + \frac{1}{2}$$

$$\geq 1 + \left(\frac{m+1}{2}\right).$$

$\therefore$ By the principle of mathematical induction, $H_{2^n} \geq 1 + \dfrac{n}{2}$.

28. Using mathematical induction, prove that

$$1^2 + 3^2 + 5^2 + \cdots + (2n - 1)^2 = \frac{n(2n - 1)(2n + 1)}{3}.$$

Solution.

Let $P(n) : 1^2 + 3^2 + 5^2 + \cdots + (2n - 1)^2 = \dfrac{n(2n - 1)(2n + 1)}{3}$.

(1) $P(1) : 1^2 = \dfrac{1(2 - 1)(2 + 1)}{3} = 1$ is true.

(2) Assume $P(m)$ is true.

That is, $P(n) : 1^2 + 3^2 + 5^2 + \cdots + (2m - 1)^2 = \dfrac{m(2m - 1)(2m + 1)}{3}$.

(3) Now,

$$1^2 + 3^2 + 5^2 + \cdots + (2m-1)^2 + [(2(m+1)-1]^2$$

$$= \frac{m(2m-1)(2m+1)}{3} + (2m+1)^2$$

$$= \frac{1}{3}[m(2m-1)(2m+1) + 3(2m+1)]^2$$

$$= \frac{2m+1}{3}[m(2m-1) + 3(2m+1)]$$

$$= \frac{2m+1}{3}[2m^2 - m + 6m + 3]$$

$$= \frac{2m+1}{3}(2m^2 + 5m + 3)$$

$$= \frac{2m+1}{3}(2m^2 + 2m + 3m + 3)$$

$$= \frac{2m+1}{3}[2m(m+1) + 3(m+1)]$$

$$= \frac{2m+1}{3}(2m+3)(m+1)$$

$$= \frac{1}{3}(m+1)(2m+1)(2m+3)$$

$$= \frac{1}{3}(m+1)[2(m+1) - 1][2(m+1) + 1].$$

∴ By the principle of mathematical induction,

$$1^2 + 3^2 + 5^2 + \cdots + (2n-1)^2 = \frac{n(2n-1)(2n+1)}{3}.$$

29. Use mathematical induction to show that $n^3 - n$ is divisible by 3, for $n \in \mathbb{N}$.

Solution.
Let $P(n) : n^3 - n$ be divisible by 3.

(1) $P(1) : 1^3 - 1 = 0$ is divisible by 3.
(2) Assume $P(m)$ is true.
   That is, $m^3 - m$ is divisible by 3.
(3) Now,

$$(m+1)^3 - (m+1) = m^3 + 3m^2 + 3m + 1 - m - 1$$

$$= m^3 + 3m^2 + 2m$$

$$= m^3 - m + 3m^2 + 2m + m$$

$$= (m^3 - m) + 3(m^2 + m).$$

Since both $m^3 - m$ and $3(m^2 + m)$ are divisible by 3, $(m + 1)^3 - (m + 1)$ is also divisible by 3.

Hence, by mathematical induction, $n^3 - n$ is divisible by 3.

2.2.4   Problems for Practice

Using the principle of mathematical induction,

1. Prove that $2 + 2^2 + 2^3 + \cdots + 2^n = 2^{n+1} - 2$.

2. Show that the sum of first n even integers is $n^2 + n$.

3. Prove that $1^2 + 3^2 + 5^2 + \cdots + (2n - 1)^2 = \dfrac{4n^3 - n}{3}$.

4. Prove that $\dfrac{1}{2} + \dfrac{1}{2^2} + \dfrac{1}{2^3} + \cdots + \dfrac{1}{2^n} = 1 - \dfrac{1}{2^n}$.

5. Show that $n^4 - 4n^2$ is divisible by 3 for all $n \in \mathbb{N}$.

6. Prove that $1 \cdot 2 + 2 \cdot 3 + \cdots + n(n + 1) = \dfrac{n(n + 1)(n + 2)}{3}$.

7. Prove that $1 + 4 + 7 + \cdots + (3n - 2) = \dfrac{n(3n - 1)}{2}$.

8. Show that $\dfrac{1}{1 \cdot 3} + \dfrac{1}{3 \cdot 5} + \dfrac{1}{5 \cdot 7} + \cdots + \dfrac{1}{(2n - 1)(2n + 1)} = \dfrac{n}{2n + 1}$.

9. Prove that the sum of the cubes of three consecutive integers is divisible by 9.

10. Show that $2^{2n} - 1$ is divisible by 3 for all $n \in \mathbb{N}$.

11. Prove that $5^n - 4n - 1$ is exactly divisible by 16 for $n \in \mathbb{N}$.

12. Show that $11^n - 4^n$ is divisible by 7 for $n \in \mathbb{N}$.

13. Show that $n + 1 < n^2$ for $n \geq 2$.

14. Show that $2n < 3^n$ for all $n \in \mathbb{N}$.

15. Show that $2^n < n^3$ for $n \geq 10$.

16. Prove that $n^3 - n$ is divisible by 6.

17. Show that $5^{2n} - 2^{5n}$ is divisible by 7.

18. Prove that $n^5 - n$ is divisible by 5.

19. Show that $10^{n+1} + 10^n + 1$ is divisible by 3.

20. Show that $n^2 < 2^n$ for $n \geq 4$.

21. Show that $2^n \geq (2n + 1)$ for $n \geq 3$.

22. Prove that

$$1 \cdot 2 \cdot 3 + 2 \cdot 3 \cdot 4 + 3 \cdot 4 \cdot 5 + \cdots + n(n + 1)(n + 2)$$
$$= \dfrac{n(n + 1)(n + 2)(n + 3)}{4}, \quad n \geq 1.$$

2.2.5 Strong Induction

Strong induction is another form of mathematical induction which is very useful in proofs. In this form, we use the basic step similar to the principle of mathematical induction, and for inductive step, we assume that $P(k)$ is true for $k = 1, 2, \ldots, m$ and then show that $P(m + 1)$ is true based on this assumption. Strong induction is also called as the second principle of mathematical induction.

Procedure for strong induction
Basis step: $P(1)$ has to be proved as true.
Inductive step: $[P(1) \wedge P(2) \wedge \cdots \wedge P(m)] \to P(m + 1)$ has to be proved as true by assuming $P(k)$ is true for $k = 1, 2, \ldots, m$.

Solved Example:
Show that if n is an integer greater than 1, then n can be written as the product of primes.

Solution.
Let $P(n) : n$ be written as the product of primes.
    Basic step: $P(2)$ is true since $2 = 1 \times 2$, product of primes.
    Inductive step:
    Assume that $P(k)$ is true for all positive integers k with $k \leq m$. To complete the inductive step, it must be shown that $P(m + 1)$ is true under this assumption.
    Two cases arise, namely

(i) when $(m + 1)$ is prime

(ii) when $(m + 1)$ is composite.

Case (i) : If $(m + 1)$ is prime, it is obvious that $P(m + 1)$ is true.

Case (ii) : If $(m + 1)$ is composite, then it can be written as a product of two positive integers a and b with $2 \leq a < b \leq m + 1$. By the induction hypothesis, both a and b can be written as the product of primes. Thus, if $(m + 1)$ is composite, it can be written as the product of primes, namely those primes in the factorisation of a and those in the factorisation of b.

2.2.6 Well-Ordering Property

The validity of mathematical induction follows from the following fundamental axioms about the set of integers.
    Every non-empty set of non-negative integers has a least element.
    The well-ordering property can often be used directly in the proof.

Solved Example:
What is wrong with this "proof" by strong induction?

Theorem: For every non-negative integer n, $5n = 0$.

Proof.
Basis step: $5 \cdot 0 = 0$.
Induction step: Suppose that $5j = 0$ for all non-negative integers j with $0 \le j \le m$. Write $m + 1 = i + j$ where i and j are natural numbers less than $m + 1$. By the induction hypothesis,

$$5(m + 1) = 5(i + 1) = 5i + 5j = 0 + 0 = 0.$$

2.3 Pigeonhole Principle

If n pigeonholes are occupied by $n + 1$ or more pigeons, then at least one pigeonhole is occupied by more than one pigeon.

Example 1. Suppose a department contains 13 professors. Then, two of the professors (pigeons) were born in the same month (pigeonhole).

Example 2. Suppose a laundry bag contains many red, white, and blue socks. Then, one needs to only grab four socks (pigeons) to be sure of getting a pair with the same colour (pigeonhole).

Example 3. Find the minimum number of elements that one needs to take from the set $S = \{1, 2, \ldots, 9\}$ to be sure that two of the numbers add up to 10. Hence, the pigeonholes are the five sets $\{1, 9\}$, $\{2, 8\}$, $\{3, 7\}$, $\{4, 6\}$, $\{5\}$. Thus, any choice of six elements (pigeons) of S will guarantee that two of the numbers add up to 10.

2.3.1 Generalized Pigeonhole Principle

If n pigeonholes are occupied by $kn + 1$ or more pigeons, where k is a positive integer, then at least one pigeonhole is occupied by $k + 1$ or more pigeons.

2.3.2 Solved Problems

1. Find the minimum number of students in a class to be sure that three of them were born in the same month.

Solution.

Here, $n = 12$ months are the pigeonholes and $k + 1 = 3$ or $k = 2$. Hence, among any $kn + 1 = 25$ students (pigeons), three of them were born in the same month.

2. Suppose a laundry bag contains many red, white, and blue socks. Find the minimum number of socks that one needs to choose in order to get two pairs (four socks) of the same colour.

    Solution.

    There are $n = 3$ colours (pigeonholes) and $k + 1 = 4$ or $k = 3$. Thus, among any $kn + 1 = 10$ socks (pigeons), four of them have the same colour.

3. Assume there are n distinct pairs of shoes in a closet. Show that if you choose $n + 1$ single shoes at random from the closet, you are certain to have a pair.

    Solution.

    The n distinct pairs constitute n pigeonholes. The $n+1$ single shoes correspond to $n + 1$ pigeons. Therefore, there must be at least one pigeonhole with two shoes, and thus you will certainly have drawn at least one pair of shoes.

4. Assume there are three men and five women at a party. Show that if these people are lined up in a row, at least two women will be next to each other.

    Solution.

    Consider the case where the men are placed so that no two men are next to each other and not at either end of the line. In this case, the three men generate four potential locations (pigeonholes) to place women (at either end of the line and two locations between men within the line). Since there are five women (pigeons), at least one slot will contain two women who must, therefore, be next to each other. If the men are allowed to be placed next to each other or at the end of the line, there are even fewer pigeonholes and, once again, at least two women will have to be placed next to each other.

5. Find the minimum number of students needed to guarantee that five of them belong to the same class (Freshman, Sophomore, Junior, Senior).

    Solution.

    Here, $n = 4$ classes are the pigeonholes and $k+1 = 5$ or $k = 4$. Thus, among any $kn + 1 = 17$ students (pigeons), five of them belong to the same class.

6. A student must take five classes from three areas of study. Numerous classes are offered in each discipline, but the student cannot take

more than two classes in any given area. Using pigeonhole principle, show that the student will take at least two classes in one area.

Solution.

The three areas are the pigeonholes, and the student must take five classes (pigeons). Hence, the student must take at least two classes in one area.

7. Let L be a list (not necessarily in alphabetical order) of the 26 letters in the English alphabet (which consists of 5 vowels, A, E, I, O, U, and 21 consonants).

   (i) Show that L has a sublist consisting of four or more consecutive consonants.

   (ii) Assuming L begins with a vowel, say A, show that L has a sublist consisting of five or more consecutive consonants.

Solution.

   (i) The five letters partition L into $n = 6$ sublists (pigeonholes) of consecutive consonants. Here, $k + 1 = 4$ and so $k = 3$. Hence, $nk + 1 = 6(3) + 1 = 19 < 21$. Hence, some sublist has at least four consecutive consonants.

   (ii) Since L begins with a vowel, the remaining vowels partition L into $n = 5$ sublists. Here, $k + 1 = 5$ and so $k = 4$. Hence, $kn + 1 = 21$. Thus, some sublist has at least five consecutive consonants.

8. Find the minimum number n of integers to be selected from $S = \{1, 2, \ldots, 9\}$ so that

   (i) the sum of two of the n integers is even

   (ii) the difference of two of the n integers is 5.

Solution.

   (i) The sum of two even integers or of two odd integers is even. Consider the subsets $\{1, 3, 5, 7, 9\}$ and $\{2, 4, 6, 8\}$ of S as pigeonholes. Hence, $n = 3$.

   (ii) Consider the five subsets $\{1, 6\}$, $\{2, 7\}$, $\{3, 8\}$, $\{4, 9\}$, $\{5\}$ of S as pigeonholes. Then, $n = 6$ will guarantee that two integers will belong to one of the subsets and their difference will be 5.

2.3.3   Another Form of Generalized Pigeonhole Principle

If m pigeons occupy n holes $(m > n)$, then at least one hole has more than $\left\lfloor \dfrac{m-1}{n} \right\rfloor + 1$ pigeons.

Here, $[x]$ denotes the greatest integer less than or equal to x, which is a real number.

2.3.4 Solved Problems

1. Show that among 100 people, at least nine of them were born in the same month.

   Solution.
   Here, number of pigeons $= m =$ number of people $= 100$.

   number of holes $= n =$ number of months $= 12$.

   Then by generalized pigeonhole principle, at least
   $$\left\lceil \frac{m-1}{n} \right\rceil + 1 = \left\lceil \frac{100-1}{12} \right\rceil + 1 = \left\lceil \frac{99}{2} \right\rceil + 1 = 8 + 1 = 9 \text{ were born}$$
   in the same month.

2. Show that if seven colours are used to paint 50 bicycles, at least eight bicycles will be the same colour.

   Solution.
   Here, number of pigeons $= m =$ number of bicycles $= 50$.

   number of holes $= n =$ number of colours $= 7$.

   Then by generalized pigeonhole principle, at least
   $$\left\lceil \frac{m-1}{n} \right\rceil + 1 = \left\lceil \frac{50-1}{7} \right\rceil + 1 = 7 + 1 = 8 \text{ bicycles will have the}$$
   same colour.

3. Show that if 25 dictionaries in a library contain a total of 40235 pages, then one of the dictionaries must have at least 1,614 pages.

   Solution.
   Here, number of pigeons $= m =$ number of bicycles $= 50$.

   number of holes $= n =$ number of colours $= 7$.

   Then by generalized pigeonhole principle, at least
   $$\left\lceil \frac{m-1}{n} \right\rceil + 1 = \left\lceil \frac{40325-1}{25} \right\rceil + 1 = \left\lceil \frac{40324}{25} \right\rceil + 1 = 1613 + 1 = 1614$$
   pages.

4. Prove that in any group of six people, there must be at least three mutual friends or at least three mutual enemies.

   Solution.
   Let those six people be A, B, C, D, E and F. Fix A. The remaining five people can be accommodated into two groups, namely

   (i) Friends of A and

   (ii) Enemies of A.

   Now, by generalized pigeonhole principle, at least one of the groups must contain $\left\lceil \frac{5-1}{2} \right\rceil + 1 = 3$ people.

    (i) If any two of these three people (B, C, D) are friends, then these two together with A form three mutual friends.

    (ii) If no two of these three people are friends, then these three people (B, C, D) are mutual enemies. In either case, we get the required conclusion.

If the group of enemies of A contains three people, by the above similar argument, we get the required conclusion.

5. If we select ten points in the interior of an equilateral triangle of side 1, show that there must be at least two points whose distance apart is less than $\frac{1}{3}$.

Solution.

Let ABC be the given equilateral triangle. Let D and E be the points of trisection of the side AB, F and G be the points of trisection of the side BC, and H and I be the points of trisection of AC, so that the triangle ABC is divided into nine equilateral triangles each of side $\frac{1}{3}$.

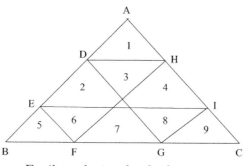

Equilateral triangle of side 1 unit

Here, number of pigeons $= m =$ number of interior points $= 10$.

    number of holes $= n =$ number of triangles $= 9$.

Then by generalized pigeonhole principle, at least one triangle contains

$$\left\lceil \frac{10 - 1}{2} \right\rceil + 1 = 2 \text{ interior points.}$$

Since each triangle is of length $\frac{1}{3}$, the distance between any two interior points of any sub-triangle cannot exceed $\frac{1}{3}$.

6. Find the minimum number of students needed to guarantee that five of them belong to the same subject, having majors as English, Maths, Physics, and Chemistry.

Solution.

Number of pigeonholes = Number of subjects = $n = 4$

Let k be the number of students (pigeons) in each subject.

Now, $k + 1 = 5 \Rightarrow k = 4$.

Therefore, the total number of students = $kn + 1 = 4(4) + 1 = 17$.

7. Show that if any 11 numbers from 1 to 20 are chosen, then 2 of them will add up to 21.

Solution.

Construct the following sets with two numbers that add up to 21.

$$A_1 = \{1, 20\}, \quad A_2 = \{2, 19\}, \quad A_3 = \{3, 18\}, \quad A_4 = \{4, 17\},$$
$$A_5 = \{5, 16\}, \quad A_6 = \{6, 15\}, \quad A_7 = \{7, 14\}, \quad A_8 = \{8, 13\},$$
$$A_9 = \{9, 12\}, \quad A_{10} = \{10, 11\}.$$

By pigeonhole principle, if any 11 numbers from 1 to 20 are chosen, then we must have to select all 2 elements from at least 1 set from the above 10 sets, which will give the sum as 21.

8. If we select any group of 1,000 students on campus, show that at least 3 of them must have the same birthday.

Solution.

The maximum number of days in a year is 366.

Here, number of students = number of pigeons = $m = 1,000$.

   Number of days in a year = number of holes = $n = 366$.

By generalized pigeonhole principle, at least

$$\left\lceil \frac{m-1}{n} \right\rceil + 1 = \left\lceil \frac{1000-1}{366} \right\rceil + 1 = 2 + 1 = 3 \text{ students must have the}$$
same birthday.

9. How many students must be in a class to guarantee that at least two students receive the same score on the final exam, if exam is graded on a scale from 0 to 100 points.

Solution.

There are 101 possible scores as $0, 1, 2, \ldots, 100$. By pigeonhole principle, we have 102 students. Hence, there must be at least two students with the same score.

Therefore, the class must contain minimum 102 students.

10. Show that among $(n + 1)$ positive integers not exceeding $2n$, there must be an integer that divides one of the other integers.

Solution.

Let the $(n + 1)$ integers be $a_1, a_2, \ldots, a_{n+1}$.

Each of these numbers can be expressed as an odd multiple of a power of 2.

That is, $\qquad\qquad a_i = 2^{k_i} \times m_i,$

where k_i is a non-negative integer, m_i is an odd number; $i = 1, 2, \ldots, n+1$.

Here, pigeon = odd positive integers $m_1, m_2, \ldots, m_{n+1}$ less than $2n$.

$\qquad$ Pigeonhole = n odd positive integers less than $2n$.

Therefore, by pigeonhole principle, two of the integers must be equal. Let it be $m_i = m_j$.

Now, $\qquad\qquad a_i = 2^{k_i} \qquad$ and $\qquad a_j = 2^{k_j} m_j$

$$\Rightarrow \frac{a_i}{a_j} = \frac{2^{k_i}}{2^{k_j}} \quad (\because m_i = m_j).$$

Case (i): If $k_i < k_j$, then 2^{k_i} divides 2^{k_j}, and hence a_i divides a_j.

Case(ii): If $k_i > k_j$, then a_j divides a_i.

11. Prove that in an equilateral triangle whose sides are of length 1 unit, if any five points are chosen, then at least two of them lie in a triangle whose sides apart is less than $\dfrac{1}{2}$.

Solution.

Let D, E, and F be midpoints of the sides AB, BC, and AC, respectively, so that triangle ABC is divided into four equilateral triangles each of side $\dfrac{1}{2}$.

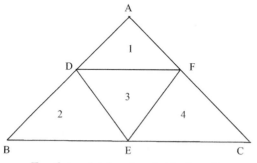

Equilateral triangle of side 1 unit

Now, number of pigeon $= m =$ number of interior points $= 5$.

Number of pigeonholes $= n =$ number of triangles $= 4$.

By pigeonhole principle, at least one triangle has more than one point (or maximum two points).

Since each triangle side is $\dfrac{1}{2}$, the distance between two interior points of any subtriangle is less than $\dfrac{1}{2}$.

12. Show that among 13 children, there are at least 2 children who were born in the same month.

    Solution.
    Assume the 13 children as pigeons and 12 months (from January to December) as the pigeonholes. Then, by the pigeonhole principle, there will be at least two children who were born in the same month.

13. Show that if any four numbers from 1 to 6 are chosen, then two of them will add up to 7.

    Solution.
    The following sets contain two numbers whose sum is 7.

    $$A_1 = \{1,6\}, \quad A_2 = \{2,5\}, \quad A_3 = \{3,4\}.$$

    The numbers from 1 to 6 can be splitted into 3 sets above who sum add up to 7. Hence if any four numbers from 1 to 6 are chosen, then two of them will belong to any one of the above 3 sets whose sum is 7.

14. Show that among any group of five (not necessarily consecutive) integers, there are two with the same remainder when divided by 4.

    Solution.
    Take any group of five integers. When these are divided by 4, each has some remainder. Since there are five integers and four possible remainders when an integer is divided by 4, the pigeonhole principle implies that given five integers, at least two have the same remainder.

15. A bag contains 12 pairs of socks (each pair is in different colour). If a person draws the socks one by one at random, determine at most how many draws are required to get at least one pair of matched socks.

    Solution.
    Let n denote the number of draws. For $n \leq 12$, it is possible that the socks drawn are of different colours, since there are 12 colours. For $n = 13$, all socks cannot have different colours, and at least two must have the same colour. Here 13 is the number of pigeons and

12 colours are 12 pigeonholes. Hence, at most 13 draws are required to have at least one pair of socks of the same colour.

16. Show that for every integer n there is a multiple of n that has only 0's and 1's in its decimal expansion.

    Solution.
    Let n be a positive integer.

    Consider the $n + 1$ integers $1, 11, 111, \ldots$ There are n possible remainders when an integer is divided by n. Since there are $n + 1$ integers in this list, by the pigeonhole principle, there must be two with the same remainder when divided by n.

    The larger number of these integers minus the smaller one is a multiple of n, which has a decimal expansion consisting entirely of 0's and 1's.

17. Prove the statement: If $m = k_{n+1}$ pigeons (where $k \geq 1$) occupy n pigeonholes, then at least one pigeonhole must contain $k + 1$ or more pigeons.

    Solution.
    Let us assume that the conclusion of the given statement is false.

    Then, every pigeonhole contains k or less number of pigeons. Then, the total number of pigeons would be nk. This is a contradiction. Hence, the assumption made is wrong, and the given statement is true.

18. Let $n_1, n_2, \ldots, n_r$ be r objects. Show that if $n_1 + n_2 + \cdots + n_r - r + 1$ objects are placed in r boxes, then for some $i = 1, 2, \ldots, r$, the ith box contains at least n_i objects.

    Solution.
    Assume that the conclusion part of the given statement is false.

    Here $n_1, n_2, \ldots, n_r$ are pigeons, r boxes are pigeonholes. Then, for holes containing n_{j-1} or less number of pigeons, $j = 1, 2, \ldots, m$. Then, the total number of pigeons would be less than or equal to

    $$(n_1 - 1) + (n_2 - 1) + \cdots + (n_r - 1) = n_1 + n_2 + \cdots + n_r - r = m - 1.$$

    This is a contradiction, since the number of pigeons is equal to m. Hence, the assumption made is wrong, and the given statement is true.

19. Seven members of a family have totally Rs. 2886 in their pockets. Show that at least one of them must have at least Rs. 416 in his pocket.

    Solution.
    Assume "members" as pigeonholes and "rupees" as pigeons.

    2886 pigeons are to be assigned to seven pigeonholes.

By generalized pigeonhole principle, at least

$$\left\lceil \frac{m-1}{n} \right\rceil + 1 = \left\lceil \frac{2886-1}{7} \right\rceil + 1 = 416 \quad \text{rupees in one member's}$$

pocket.

20. If nine books are to be kept in four shelves, there must be at least one shelf which contains at least three books.

    Solution.

    Assume "books" as pigeons and "shelves" as pigeonholes.

    Nine pigeons are to be assigned to four shelves.

    By generalized pigeonhole principle, at least

    $$\left\lceil \frac{m-1}{n} \right\rceil + 1 = \left\lceil \frac{9-1}{4} \right\rceil + 1 = 3 \text{ books in one shelf.}$$

21. How many people must you have to guarantee that at least nine of them will have birthdays in the same day of the week.

    Solution.

    Assume "days in a week" as pigeonholes and "people" as pigeons.

    We have to find the number of people (pigeons) to be assigned to seven pigeonholes.

    By generalized pigeonhole principle (given at least nine of them will have birthdays in the same week),

    $$\left\lceil \frac{m-1}{n} \right\rceil + 1 = 9$$

    $$\left\lceil \frac{m-1}{7} \right\rceil + 1 = 9$$

    $$\frac{m-1+7}{7} = 9$$

    $$\frac{m+6}{7} = 9$$

    $$m = 57.$$

    Hence, there must be 57 people to guarantee that at least nine of them will have birthdays in the same day of the week.

22. Show that if 30 dictionaries in a library contain a total of 61327 pages, then one of the dictionaries must have at least 2045 pages.

    Solution.

    Assume "pages" as pigeons and "dictionaries" as pigeonholes.

    61327 pages (pigeons) are to be assigned to 30 dictionaries (pigeonholes).

    By generalized pigeonhole principle, one dictionary must contain

$$\left\lceil \frac{m-1}{n} \right\rceil + 1 = \left\lceil \frac{61327 - 1}{30} \right\rceil + 1 = 2045 \text{ pages.}$$

23. What is the maximum number of students required in a mathematics class to be sure that at least six will receive the same grade, if there are five possible grades A, B, C, D, and F?

    Solution.
    The minimum number of students needed to ensure that at least six students receive the same grade is the smaller integer N such that $\left\lceil \dfrac{N}{5} \right\rceil = 6$. The smallest such integer is $N = 5 \cdot 5 + 1 = 26$.
    If you have only 25 students, it is possible for there to be five who have received each grade so that no six students have received the same grade.

    Therefore, 26 is the minimum number of students needed to ensure that at least six students will receive the same grade.

24. How many persons must be chosen in order that at least five of them will have birthdays in the same calendar month?

    Solution.
    Let n be the required number of persons. Since the number of months over which the birthdays are distributed is 12, the minimum number of persons who have their birthdays in the same month is, by the generalized pigeonhole principle, equal to $\left\lceil \dfrac{m-1}{12} \right\rceil + 1$. This number is 5.
    That is, $\left\lceil \dfrac{m-1}{12} \right\rceil + 1 = 5$ or $m = 49$.
    Hence, the number of persons is at least 49.

25. Find the least number of ways of choosing three different numbers from 1 to 10 so that all choices have the same sum.

    Solution.
    From the numbers 1 to 10, we can choose three different numbers in $10C_3 = 120$ ways.
    The smallest possible sum that we get from a choice is $1 + 2 + 3 = 6$, and the largest sum is $8 + 9 + 10 = 27$. Thus, the sums vary from 6 to 27 (both inclusive), and these sums are 22 in number.
    Accordingly, there are 120 choices (pigeons) and 22 sums (pigeonholes).
    Therefore, the least number of choices assigned to the same sum is, by the generalized pigeonhole principle, $\left\lceil \dfrac{120 - 1}{22} \right\rceil + 1 = 6$.

26. Show that if any five numbers from 1 to 8 are chosen, then two of them will have their sum equal to 9.

Solution.

Consider the following sets:

$$A_1 = \{1, 8\}, \quad A_2 = \{2, 7\}, \quad A_3 = \{3, 6\}, \quad A_4 = \{4, 5\}.$$

These are the only sets containing two numbers from 1 to 8, whose sum is 9.

Since every number from 1 to 8 belongs to one of the above sets, each of the five numbers chosen must belong to one of the sets.

Since there are only four sets, two of the five chosen numbers have to belong to the same set (by the pigeonhole principle).

These two numbers have their sum equal to 9.

2.3.5 Problems for Practice

1. If m is an odd positive integer, then prove that there exists a positive integer n such that m divides $2^n - 1$.

2. A man hiked for 10 hours and covered a total distance of 45 km. It is known that he hiked 6 km in the first hour and only 3 km in the last hour. Show that he must have hiked at least 9 km within a certain period of two consecutive hours.

3. Consider a tournament in which each of n players plays against every other player and each player wins at least once. Show that there are at least two players having the same number of wins.

4. Show that any set of seven distinct integers includes two integers, x and y, such that either $x + y$ or $x - y$ is divisible by 10.

5. What is the minimum number of students, each of whom comes from one of the 50 states, who must be enrolled in a university to guarantee that there are at least 100 who come from the same state?

6. Show that if any eight positive integers are chosen, two of them will have the same remainder when divided by 7.

7. A drawer contains a dozen brown socks and a dozen black socks, all unmatched. A man takes socks out at random in the dark.

   (i) How many socks must he take out to be sure that he has at least two socks of the same colour?

   (ii) How many socks must he take out to be sure that he has at least two black socks?

8. There are 38 different time periods during which classes at a university can be scheduled. If there are 677 different classes, how many different rooms will be needed?

9. Construct a sequence of 16 positive integers that has no increasing or decreasing subsequence of five terms.

10. Suppose there are 26 students and seven cars to transport them. Then, show that at least one car must have four or more passengers.

11. Show that in any set of eleven integers, there are two whose difference is divisible by 15.

12. Show that in any room of people who have been doing handshaking, there will always be at least two people who have shaken hands the same number of times.

13. Show that if nine colours are used to paint 100 houses, at least 12 houses will be of the same colour.

14. Show that if any five integers from 1 to 8 are chosen, then at least two of them will have a sum 9.

15. Prove that if any 30 people are selected, then we may choose a subset of five so that all five were born on the same day of the week.

16. Show that in any set of 11 integers, there are two whose difference is divisible by 15.

17. A drawer contains ten black and ten white socks. What is the last number of socks one must pull out to be sure to get a matched pair?

18. In a group of 13 children, show that there must be at least two children who were born in the same month.

19. Prove that every set of 37 positive integers contains at least two integers that leave the same remainder upon division by 36.

20. Let A be some fixed ten element set of $\{1, 2, 3, \ldots, 50\}$. Show that A possesses two different five element subsets, the sum of whose elements are equal.

2.4 Permutation

Any arrangement of a set of n objects in a given order is called a permutation of the objects (taken all at a time). An arrangement of any $r \leq n$ of these objects in a given order is called an r-permutation or a permutation of the n objects taken r at a time.

For example, consider the set of letters: a, b, c, and d. Then,

(i) *bdca*, *dcba*, and *acdb* are permutations of the four letters (taken all at a time).

(ii) *bad adb*, *cbd*, and *bca* are permutations of the four letters taken three at a time.

(iii) *ad, cb, da,* and *bd* are permutations of the four letters taken two at a time.

The number of permutations of n objects taken r at a time is denoted by

$$nP_r \quad \text{or} \quad P(n,r) \quad \text{or} \quad P_{n,r} \quad \text{or} \quad P_r^n \quad \text{or} \quad (n)_r.$$

We shall use nP_r or $P(n,r)$.

Example:

Find the number of permutations of six objects, say, A, B, C, D, E, and F taken three at a time. In other words, find the number of three-letter words using only the given six letters without repetition.

Solution.

Let the general three-letter words be represented by the following three boxes:

$$\square \quad \square \quad \square$$

The first letter can be chosen in six different ways. Following this, the second letter can be chosen in five different ways, and, following this, the last letter can be chosen in four different ways. Write each number in its appropriate box as follows:

$$\boxed{6} \;\; \boxed{5} \;\; \boxed{4}$$

Therefore, by the fundamental principle of counting, there are $6 \times 5 \times 4 = 120$ possible three-letter words without repetition from the six letters, or there are 120 permutations of six objects taken three at a time.

$$6P_3 = P(6,3) = 120.$$

Formula for nP_r:

$$nP_r = \frac{n!}{(n-r)!}$$

Remark:

(i) When $r = n$, then $nP_n = n!$

(ii) There are $n!$ permutations of n objects (taken all at a time). For example, there are $3! = 1 \times 2 \times 3 = 6$ permutations of the three letters $a, b,$ and c. They are $abc, acb, bac, bca, cab,$ and cba.

2.4.1   Permutations with Repetitions

The number of permutations of n objects of which n_1 is alike, n_2 is alike,..., n_r is alike is

$$P(n; n_1, n_2, \ldots, n_r) = \frac{n!}{n_1! n_2! \ldots n_r!}.$$

2.4.2   Solved Problems

1. How many seven-letter words can be formed using the letters of the word "BENZENE"?

   Solution.

   There are three E's and two N's in the given word.

   Therefore, here $n = 7$, $n_1 = 3$, $n_2 = 2$.

   $$P(n; n_1, n_2) = \frac{n!}{n_1! n_2!}$$

   $$P(7; 3, 2) = \frac{7!}{3! \times 2!} = \frac{7 \times 6 \times 5 \times 4 \times 3 \times 2 \times 1}{6 \times 2} = 420.$$

2. How many different signals, each consisting of eight flags hung in a vertical line, can be formed from a set of four indistinguishable red flags, three indistinguishable white flags, and a blue flag?

   Solution.

   Here $n = 8$, $n_1 = 4$, $n_2 = 3$.

   $$P(n; n_1, n_2) = \frac{n!}{n_1! n_2!}$$

   $$P(8; 4, 3) = \frac{8!}{4! \times 3!} = 280.$$

3. There are four bus lines between A and B, and three bus lines between B and C. In how many ways can a man travel

   (i) by bus from A to C by way of B?

   (ii) round-trip by bus from A to C by way of B?

   (iii) round-trip by bus from A to C by way of B, if he does not want to use a bus line more than once?

   Solution.

   (i) There are four ways to go from A to B and three ways to go from B to C. Hence, there are $4 \times 3 = 12$ ways to go from A to C by way of B.

   (ii) There are 12 ways to go from A to C by way of B and 12 ways to return. Hence, there are $12 \times 12 = 144$ ways to travel round-trip.

   (iii) The men will travel from A to B to C to B to A. Enter these letters connecting arrows as follows:

   $$A \longrightarrow B \longrightarrow C \longrightarrow B \longrightarrow A.$$

   The man can travel four ways from A to B and three ways from B to C, but he can only travel two ways from C to B and three ways from B to A since he does not want to use a bus line more than once. Enter these numbers above the corresponding arrows as follows:

$$A \xrightarrow{4} B \xrightarrow{3} C \xrightarrow{2} B \xrightarrow{3} A.$$

Therefore, there are $4 \times 3 \times 2 \times 3 = 72$ ways to travel round-trip without using the same bus line more than once.

4. Suppose repetitions are not permitted.

   (i) How many three-digit numbers can be formed from the six digits $2, 3, 5, 6, 7,$ and 9?

   (ii) How many of these numbers are less than 400?

   (iii) How many are even?

   Solution.
   In each case, draw three boxes ☐ ☐ ☐ to represent an arbitrary number, and then write in each box the number of digits that can be placed there.

   (i) The box on the left can be filled in six ways. Following this, the middle box can be filled in five ways. Lastly, the box on the right can be filled in four ways: | 6 | 5 | 4 |. Therefore, there are $6 \times 5 \times 4 = 120$ numbers.

   (ii) The box on the left can be filled only in two ways by 2 or 3, since each number must be less than 400. The middle box can be filled in five ways. Lastly, the box on the right can be filled in four ways. Therefore, there are $2 \times 5 \times 4 = 40$ numbers.

   (iii) The box on the right can be filled in only two ways by two or six, since the numbers must be even. The box on the left can be filled in five ways, and lastly, the middle box can be filled in four ways: | 5 | 4 | 2 |. Therefore, there are $5 \times 4 \times 2 = 40$ numbers.

5. Find the number of ways in which a party of seven persons can arrange themselves:

   (i) in a row of seven chairs

   (ii) around a circular table.

   Solution.

   (i) The seven persons can arrange themselves in a row in $7 \times 6 \times 5 \times 4 \times 3 \times 2 \times 1 = 7!$ ways.

   (ii) One person can sit at any place in the circular table. The other six persons can then arrange themselves in $6 \times 5 \times 4 \times 3 \times 2 \times 1 = 6!$ ways around the table.

Remark. This is an example of a circular permutation. In general, n objects can be arranged in a circle in
$$(n-1) \times (n-2) \times \cdots \times 3 \times 2 \times 1 = (n-1)! \text{ ways.}$$

6. Find the number of distinct permutations that can be formed from all the letters of the word:

   (i) RADAR
   (ii) UNUSUAL.

   Solution.

   (i) $\dfrac{5!}{2! \times 2!} = 30$, since there are five letters of which two are R and two are A.

   (ii) $\dfrac{7!}{3!} = 840$, since there are seven letters of which three are U.

7. In how many ways can four mathematics books, three history books, three chemistry books, and two sociology books be arranged on a shelf so that all books of the same subject are together?

   Solution.
   First, the books must be arranged on the shelf in four units according to subject matter: ☐☐ ☐☐ . The box on the left can be filled by any of the four subjects, the next by any three remaining subjects, the next by any two remaining subjects, and the box on the right by the last subject: | 4 | 3 | 2 | 1 | . Therefore, there are $4 \times 3 \times 2 \times 1 = 4!$ ways to arrange the books on the shelf according to subject matter.

   Now, in each of the above cases, the mathematics books can be arranged in 4! ways, the history books in 3! ways, the chemistry books in 3! ways, and the sociology books in 2! ways. Thus, altogether, there are $4! \times 4! \times 3! \times 3! \times 2! = 41472$ arrangements.

8. Find n if

   (i) $P(n,2) = 72$
   (ii) $P(n,4) = 42P(n,2)$
   (iii) $2P(n,2) + 50 = P(2n,2)$.

   Solution.

   (i) $P(n,2) = n(n-1) = n^2 - n$

$$\therefore \qquad n^2 - n = 72 \Rightarrow n^2 - n - 72 = 0$$
$$\Rightarrow (n-9)(n+8) = 0$$
$$\Rightarrow n = 9, -8$$
$$\Rightarrow n = 9 \text{ since } n \text{ is positive.}$$

(ii) $P(n,4) = n(n-1)(n-2)(n-3)$ and $P(n,2) = n(n-1)$
Therefore, $n(n-1)(n-2)(n-3) = 42n(n-1)$.
If $n \neq 0$, $n \neq 1$,

$$(n-2)(n-3) = 42$$
$$\Rightarrow n^2 - 5n + 6 = 42$$
$$\Rightarrow n^2 - 5n - 36 = 0$$
$$\Rightarrow (n-9)(n+4) = 0$$
$$\Rightarrow n = 9 \quad \text{since } n \text{ is positive.}$$

(iii) $P(n,2) = n(n-1) = n^2 - n$
$P(2n,2) = 2n(2n-1) = 4n^2 - 2n$
Therefore,

$$2(n^2 - n) + 50 = 4n^2 - 2n$$
$$\Rightarrow 2n^2 - 2n + 50 = 4n^2 - 2n$$
$$\Rightarrow 2n^2 = 50$$
$$\Rightarrow n^2 = 25$$
$$\Rightarrow n = 5 \quad \text{since } n \text{ must be positive.}$$

9. In how many ways can six persons occupy three vacant seats?

Solution.
Total number of ways $= P(6,3) = 6 \times 5 \times 4 = 120$ ways.

10. How many permutations of the letters A, B, C, D, E, F, G, H contain the string ABC?

Solution.
Since the letters A, B, and C must occur as block, we can find the answer by finding number of permutations of six objects, namely the block ABC and individual letters D, E, F, G, and H.

Therefore, there are 6! = 720 permutations of the letters A, B, C, D, E, F, G, H in which ABC occurs.

11. If $P(12,r) = 1320$, find r.

Solution.
$$P(12,r) = 12 \times 11 \times 10 \ldots r \text{ factors}$$
$$\Rightarrow 1320 = 12 \times 11 \times 10 \ldots r \text{ factors}$$
$$\Rightarrow r = 3.$$

12. In how many of the permutations of ten things taken four at a time will

(i) one thing always occur

(ii) one thing never occur.

Solution.

(i) We can keep aside the particular thing which will always occur; the number of permutations of nine things taken three at a time is $P(9, 3)$. Now, this particular thing can take up any one of the 4 places, and 50 can be arranged in four ways.
Therefore, the total number of permutations
$$= P(9, 3) \times 4 = 9 \times 8 \times 7 \times 4 = 2016.$$

(ii) If we are keeping the particular thing aside as never to occur, the number of permutations of nine things $(10 - 1 = 9)$ taken four at a time is $P(9, 4) = 9 \times 8 \times 7 \times 6 = 3026.$

13. In how many ways can six boys and four girls be arranged in a straight line so that no two girls are ever together.

Solution.

The arrangement may be done in two operations.

(i) First, we fix the positions of six boys. Their positions are indicated by $B_1, B_2, \ldots, B_6$. That is,

$$\text{X } B_1 \text{ X } B_2 \text{ X } B_3 \text{ X } B_4 \text{ X } B_5 \text{ X } B_6.$$

This can be done in 6! ways.

(ii) If the positions of girls are fixed at places including those at the two ends as shown by the crosses, the four girls will never come together. In any one of these arrangements, there are seven places for four girls, and so the girls can sit in $7P_4$ ways.
∴ The number of ways of seating six boys and four girls
$$= 7P_4 \times 6! = 7 \times 6 \times 5 \times 4 \times 6 \times 5 \times 4 \times 3 \times 2 \times 1 = 604800.$$

14. There are six books on Economics, three on Commerce and two on History. In how many ways can these be placed on a shelf of books if the same subjects are to be together?

Solution.

Six Economics books can be arranged in $6P_6$ ways or 6! ways. Three commerce books can be arranged in $3P_3$ ways or 3! ways. Two history books can be arranged in $2P_2$ ways or 2! ways.

The three subject books, Economics, Commerce, and History books, can be arranged in $3P_3$ ways or 3! ways.

∴   The total number of required arrangements
$$= 6! \times 3! \times 2! \times 3! \text{ ways} = 51840 \text{ ways}.$$

15. Suppose there are six boys and five girls.

(i) In how many ways can they sit in a row?

(ii) In how many ways can they sit in a row if the boys and girls each sit together?

(iii) In how many ways can they sit in a row if the girls are to sit together and the boys are not to sit together?

(iv) How many seating arrangements are there with no two girls sitting together?

Solution.

(i) There are $6 + 5 = 11$ persons, and they can sit in $11P_{11}$ ways. That is, $11P_{11} = 11!$ ways.

(ii) The boys among themselves can sit in 6! ways, and girls among themselves can sit in 5! ways.

They can be considered as two units and can be permuted in 2! ways.

Thus, the required seating arrangements can be done in
$$= 2! \times 6! \times 5! \text{ ways} = 2 \times 720 \times 120 \text{ ways} = 172800 \text{ ways.}$$

(iii) The boys can sit in 6! ways and girls in 5! ways. Since girls have to sit together, they are considered as one unit. Among the six boys, either 0 or 1 or 2 or 3 or 4 or 5 or 6 have to sit to the left of the girls' units. Of these seven ways, 0 and 6 cases have to be omitted as the boys do not sit together.

Thus, the required number of arrangements
$$= 5 \times 6! \times 5! \text{ ways} = 5 \times 720 \times 120 \text{ ways} = 432000 \text{ ways.}$$

(iv) The boys can sit in 6! ways. There are seven places where the girls can be placed. Thus, the total arrangements are
$$= 7P_5 \times 6! \text{ ways}$$
$$= \frac{7!}{2!} \times 720$$
$$= 2520 \times 720$$
$$= 1814400 \text{ ways.}$$

16. Find the number of ways in which five boys and five girls can be seated in a row if the boys and girls are to have alternate seats.

Solution.
Case (i): Boys can be arranged among themselves in 5! ways.

$$\square B \square B \square B \square B \square B \square$$

There are six places for girls. Hence, there are $6P_5 \times 5!$ arrangements.
Case (ii): Girls can be arranged in 5! ways.

$$\square G \square G \square G \square G \square G \square$$

There are six places for boys. Hence, there are $6P_5 \times 5!$ ways. Hence, taking two cases into account, there are $2 \times 6P_5 \times 5!$ arrangements in total.

∴ There are $2 \times 120 \times 6 = 240$ ways.

17. How many permutations of $\{a, b, c, d, e, f, g\}$

(i) end with a
(ii) begin with c

(iii) begin with c and end with a

(iv) have c and a occupying the end places?

Solution.

(i) The last position can be filled in only one way.

   The remaining six letters can be arranged in 6! ways.

   $\therefore$ The total number of permutations ending with a

   $= 6! \times 1 = 720$ ways.

(ii) The first position can be filled in only one way.

   The remaining six letters can be arranged in 6! ways.

   $\therefore$ The total number of permutations starting with c

   $= 1 \times 6!$ ways $= 720$ ways.

(iii) The first position can be filled in only one way, and the last position can be filled in only one way.

   The remaining five letters can be arranged in 5! ways.

   $\therefore$ The total number of permutations begin with c and end with a is $= 1 \times 5! \times 1$ ways $= 120$ ways.

(iv) c and a occupy end positions in 2! ways, and the remaining five letters can be arranged in 5! ways.

   $\therefore$ The total number of permutations

   $= 5! \times 2!$ ways $= 240$ ways.

18. How many bit strings of length 10 contain

   (i) exactly four 1's

   (ii) at most four 1's

   (iii) at least four 1's

   (iv) an equal number of 0's and 1's?

Solution.

(i) A bit string of length 10 can be considered to have ten positions and should be filled with four 1's and six 0's.

   $\therefore$ Required number of bit strings $= \dfrac{10!}{4! \times 6!} = 210.$

(ii) Required number of bit strings

$$= \frac{10!}{0! \times 10!} + \frac{10!}{1 \times 9!} + \frac{10!}{2! \times 8!} + \frac{10!}{3! \times 7!} + \frac{10!}{4! \times 6!} = 386.$$

(iii) Required number of bit strings

$$= \frac{10!}{4! \times 6!} + \frac{10!}{5! \times 5!} + \frac{10!}{6! \times 4!} + \frac{10!}{7! \times 3!} + \frac{10!}{8! \times 2!} + \frac{10!}{9! \times 1!} + \frac{10!}{10! \times 0!} = 848.$$

(iv) Required number of bit strings $= \dfrac{10!}{5! \times 5!} = 252.$

19. Suppose that there are 9 faculty members in the mathematics department and 11 in the computer science department. How many ways are there to select a committee to develop a discrete mathematics course in a school if the committee is to consist of three faculty members from the mathematics department and four from the computer science department?

Solution.
By the product rule, the answer is the product of the number of 3-combinations of a set with nine elements and the number of 4-combinations of a set with 11 elements. The number of ways to select the committee

$$= 9C_3 \times 11C_4 = \frac{9!}{3! \times 6!} \times \frac{11!}{4! \times 7!} = 84 \times 330 = 27720.$$

20. How many possibilities are there for the win, place, and show (first, second, and third) positions in a horse race with 12 horses if all orders of finish are possible?

Solution.
The number of ways to pick the three winners is the number of ordered selections of three elements from 12.

∴ The required number of possibilities $= 12P_3 = 12 \times 11 \times 10 = 1320.$

2.4.3 Problems for Practice

1. How many automobile license plates can be made if each plate contains two different letters followed by three different digits? Solve the problem if the first digit cannot be 0.

2. There are six roads between A and B and four roads between B and C. Find the number of ways in which one can drive

   (i) from A to C by way of B
   (ii) round-trip from A to C by way of B
   (iii) round-trip from A to C by way of B without using the same road more than once.

3. Find the number of ways in which six people can ride a toboggan if one of a subset of three must drive.

4. (i) Find the number of ways in which five persons can sit in a row.
   (ii) How many ways are there if two of the persons insist on sitting next to one another?
   (iii) Solve (i) assuming they sit around a circular table.
   (iv) Solve (ii) assuming they sit around a circular table.

5. Find the number of ways in which five large books, four medium-size books, and three small books can be placed on a shelf so that all books of the same size are together.

6. (i) Find the number of permutations that can be performed from the letters of the word ELEVEN.

   (ii) How many of them begin and end with E?

   (iii) How many of them have three E's together?

   (iv) How many begin with E and end with N?

7. (i) In how many ways can three boys and two girls sit in a row?

   (ii) In how many ways can they sit in a row if the boys and girls are each to sit together?

   (iii) In how many ways can they sit in a row if just the girls are to sit together?

8. Show that

   (i) $P(n,0) + P(n,1) + P(n,2) + \cdots + P(n,n) = 2^n$.

   (ii) $P(n,0) - P(n,1) + P(n,2) - P(n,3) + \cdots + P(n,n) = 0$.

9. How many bit strings of length 10 contain at least three 1's and at least three 0's?

10. How many ways are there for eight men and five women to stand in a line so that no two women stand next to each other?

11. The English alphabet contains 21 consonants and five vowels. How many strings of six lowercase letters of the English alphabet contain

    (i) exactly one vowel

    (ii) exactly two vowels

    (iii) at least one vowel

    (iv) at least two vowels?

12. A committee of 11 members sit at a round table. In how many ways can they be seated if the "president" and "secretary" choose to sit together?

13. In an examination, six papers are set of which two are mathematics. In how many ways can the examination be arranged if the mathematics papers are not to be together?

14. In how many ways can eight people sit around a table?

15. How many numbers are there in all which consist of five digits?

16. How many odd numbers of three digits can be formed with $1, 2, 3, 4$, and 5?

2.5 Combination

Suppose we have a collection of n objects. A combination of these n objects taken r at a time is any selection of r of the objects where order does not count. In other words, an r-combination of a set of n objects is any subset of r elements.

For example, the combinations of the letters a, b, c, d taken three at a time are

$$\{a, b, c\}, \quad \{a, b, d\}, \quad \{a, c, d\}, \quad \{b, c, d\}$$

or simply abc, abd, acd, bcd, respectively.

It can be noted that the following combinations are equal:

$$abc, acb, bac, bca, cab \quad \text{and} \quad cba.$$

That is, each denote the same set $\{a, b, c\}$.

The number of combinations of n objects taken r at a time is denoted by $C(n, r)$. The symbols nC_r, $C_{n,r}$ and C_r^n can also be used.

Formula for nC_r:

$$\boxed{nC_r = \frac{n!}{r!(n-r)!}}$$

2.5.1 Solved Problems

1. How many committees of three can be formed from eight people?

   Solution.
   Each committee is a combination of eight people taken three at a time. Therefore, the number of committees that can be formed is
   $$8C_3 = \frac{8!}{3! \times 5!} = 56.$$

2. A farmer buys three cows, two pigs, and four hens from a man who has six cows, five pigs, and eight hens. How many choices does the farmer have?

   Solution.
   The farmer can choose the cows in $6C_3$ ways, the pigs in $5C_2$ ways, and the hens in $8C_4$ ways.

   Hence, altogether he can choose the animals in
   $$6C_3 \times 5C_2 \times 8C_4 = 20 \times 10 \times 70 = 14000 \text{ ways.}$$

3. In how many ways can a committee consisting of three men and two women be chosen from seven men and five women?

   Solution.
   The three men can be chosen from the seven men in $7C_3$ ways, and the two women can be chosen from the five women in $5C_2$ ways. Hence, the committee can be chosen in $7C_3 \times 5C_2 = 350$ ways.

4. How many committees of five with a given chairperson can be selected from 12 persons?

   Solution.
   The chairperson can be chosen in 12 ways, and, following this, the other four on the committee can be chosen from the 11 remaining

in $11C_4$ ways. There are $12 \times 11C_4 = 12 \times 330 = 3960$ such committees.

5. A bag contains six white marbles and five red marbles. Find the number of ways in which four marbles can be drawn from the bag if

(i) they can be any colour

(ii) two must be white and two red

(iii) they must all be of the same colour.

Solution.

(i) The four marbles (of any colour) can be chosen from the 11 marbles in $11C_4 = 330$ ways.

(ii) The two white marbles can be chosen in $6C_2$ ways, and the two red marbles can be chosen in $5C_2$ ways. Thus, there are $6C_2 \times 5C_2 = 150$ ways of drawing two white marbles and two red marbles.

(iii) There are $6C_4 = 15$ ways of drawing four white marbles and $5C_4 = 5$ ways of drawing four red marbles. Thus, there are $15 + 5 = 20$ ways of drawing four marbles of the same colour.

6. In how many ways can a set of five letters be selected from the English alphabet?

Solution.

The number of ways to select five letters from 26 alphabets is
$$26C_5 = 65780.$$

7. How many bit strings of length n contain exactly r 1's?

Solution.

The positions of r 1's in a bit string of length n form r-combination of the set $\{1, 2, \ldots, n\}$. Hence, there are nC_r bit strings of length n that contain exactly r 1's.

Note:

(i) $nC_n = nC_0 = 1$.

(ii) $nC_r = nC_{n-r}$.

(iii) $nC_r = \frac{nP_r}{r!}$.

(iv) $nC_x = nC_y \Rightarrow n = x + y$ or $x = y$.

8. Find the value of n if $20C_{n+2} = 20C_{2n-1}$.

Solution.

$$\text{Given:} \quad 20C_{n+2} = 20C_{2n-1}$$
$$\Rightarrow \quad n + 2 = 2n - 1 \quad (\because nc_x = nc_y \Rightarrow x = y)$$
$$\Rightarrow \quad n = 3.$$

9. How many ways are there to form a committee, if the committee consists of 3 educationalists and 4 socialists, if there are 9 educationalists and 11 socialists.

Solution.

Three educationalists can be chosen from nine educationalists in $9C_3$ ways.

Four socialists can be chosen from 11 socialists in $11C_4$ ways.

Hence, by product rule, the number of ways to select the committee

$$= 9C_3 \times 11C_4$$

$$= \frac{9!}{3! \times 6!} \times \frac{11!}{4! \times 7!} = 27720 \text{ ways.}$$

10. A team of 11 players is to be chosen from 15 members. In how many ways can this be done if

   (i) one particular player is always included

   (ii) two such players have to be always included?

Solution.

   (i) Let one player be fixed. The remaining players are 14. Out of these 14 players, we have to select ten players in $14C_{10} = 1001$ ways.

   (ii) Let two players be fixed. The remaining players are 13. Out of these 13 players, we have to select nine players in $13C_9 = 715$ ways.

11. Find the number of diagonals that can be drawn by joining the angular points of a heptagon.

Solution.

A heptagon has seven angular points and seven sides. The join of two angular points is either a side or a diagonal.

The number of lines joining the angular points $= 7C_2 = \dfrac{7 \times 6}{1 \times 2} = 21$.

But the number of sides $= 7$.

$\therefore$ The number of diagonals $= 21 - 7 = 14$.

12. There are five questions in a question paper. In how many ways can a boy solve one or more questions?

Solution.

The boy can dispose of each question in two ways. He may either solve it or leave it. Thus, the number of ways of disposing all the questions $= 2^5$.

But this includes the case in which he has left all the questions unsolved.

∴   The total number of ways of solving the paper $= 2^5 - 1 = 31$.

13. Find the value of r if $20C_r = 20C_{r+2}$.

Solution.

$$20C_r = 20C_{r+2}$$
$$\Rightarrow \qquad 20C_r = 20C_{20-(r+2)}$$
$$\Rightarrow \qquad r = 20 - (r+2) \quad (\because r = r+2 \Rightarrow 2 = 0 \quad \text{is not possible})$$
$$\Rightarrow \qquad 2r = 18$$
$$\Rightarrow \qquad r = 9.$$

14. If $nC_5 = 20 \cdot nC_4$, find n.

Solution.

$$nC_5 = 20 \times C_4$$
$$\frac{n(n-1)(n-2)(n-3)(n-4)}{1 \times 2 \times 3 \times 4 \times 5} = 20 \times \frac{n(n-1)(n-2)(n-3)}{1 \times 2 \times 3 \times 4}$$
$$\frac{n-4}{5} = 20$$
$$n - 4 = 100$$
$$n = 104.$$

15. From a committee consisting of six men and seven women, in how many ways can we select a committee of

(i) three men and four women

(ii) four members that has at least one woman

(iii) four persons that has at most one man

(iv) four persons of both genders

(v) four persons in which Mr and Mrs Joseph are not included.

Solution.

(i) Three men can be selected from six men in $6C_3$ ways.
Four women can be selected from seven women in $7C_4$ ways.
∴   By product rule, the committee of three men and four women can be selected in $6C_3 \times 7C_4 = 700$ ways.

(ii) For the committee of at least one woman, we have the following possibilities:

      (i) One woman and three men

      (ii) Two women and two men

      (iii) Three women and one man

      (iv) Four women and zero men.

      Hence, the selection can be done in
$$= 7C_4 \times 6C_3 + 7C_2 \times 6C_2 + 7C_3 \times 6C_1 + 7C_4 \times 6C_0 = 700$$
ways.

(iii) For the committee of at most one man, we have the following possibilities:

      (i) One man and three women

      (ii) Zero men and four women.

      Hence, the selection can be done in
$$= 6C_1 \times 7C_3 + 6C_0 \times 7C_4 = 245 \text{ ways.}$$

(iv) For the committee of both genders, we have the following possibilities:

      (i) One man and three women

      (ii) Two men and two women

      (iii) Three men and one woman

      which can be done in $6C_1 \times 7C_3 + 6C_2 \times 7C_2 + 6C_3 \times 7C_1 = 665$ ways.

(v) Since the committee does not consist of Mr. and Mrs. Joseph, we have five men and six women in the committee.

      Now, we can select 4 members from 11 members in $11C_4 = 330$ ways.

2.5.2   Problems for Practice

1. A woman has 11 close friends.

   (i) In how many ways can she invite five of them to dinner?

   (ii) In how many ways if two of the friends are married and will not attend separately?

   (iii) In how many ways if two of them are not on speaking terms and will not attend together?

2. A woman has 11 close friends of whom six are also women.

   (i) In how many ways can she invite three or more to a party?

   (ii) In how many ways can she invite three or more of them if she wants the same number of men and women (including herself)?

3. A student is to answer 10 out of 13 questions in an exam.

   (i) How many choices does he have?

   (ii) How many if he must answer the first two questions?

    (iii) How many if he must answer the first or second question not both?

    (iv) How many if he must answer exactly three out of the first five questions?

    (v) How many if he must answer at least three of the first five questions?

4. How many diagonals are there in a polygon of ten sides?

5. A committee is to consist of two men and three women. How many different committees are possible if five men and seven women are eligible.

6. How many different groups can be selected for playing tennis out of four ladies and three gentlemen, there being one lady and one gentleman on each side?

7. From a committee of five women and seven men, in how many ways can a subcommittee of four be chosen so as to contain one particular man?

8. In how many ways can a selection be made out of five oranges, eight apples, and seven plantains?

9. In how many ways can 20 students be divided into four equal groups?

10. How many bit strings of length 10 have

    (i) exactly three 0's

    (ii) at least three 1's

    (iii) more 0's than 1's

    (iv) an odd number of 0's?

11. How many bit strings of length 12 contain

    (i) exactly three 1's

    (ii) at least three 1's

    (iii) an equal number of 1's and 0's?

12. In how many ways can a party of 16 people can be conveyed in two vehicles, one of which will not hold more than eight and the other not more than ten?

13. In how many ways can a committee of 8 be chosen from 12 socialists and 9 conservatives to give a socialist majority with at least 2 conservatives included?

14. A committee of 12 is to be selected from 10 men and 10 women. In how many ways can the selection be carried out if

    (i) there are no restrictions

    (ii) there must be equal number of men and women

(iii) there must be an even number of women

(iv) there must be more women than men

(v) there must be at least eight men?

2.5.3 Recurrence Relation

A recurrence relation for the sequence $\{f_n\}$ is a formula that expresses f_n in terms of one or more of the previous terms of the sequence, namely $f_0, f_1, \ldots, f_{n-1}$, for all integers n with $n \geq n_0$, where n_0 is non-negative integer.

A sequence is called a solution of a recurrence relation if its terms satisfy the recurrence relation.

2.5.4 Solved Problems

1. Determine whether the sequence $\{f_n\} = \{3n\}$ is a solution of the recurrence relation: $f_n = 2f_{n-1} - f_{n-2}$, for $n = 2, 3, 4, \ldots$

   Solution.
   Suppose $f_n = 3n$. Then for $n \geq 2$,

   $$\begin{aligned}
   f_n &= 2f_{n-1} - f_{n-2} \\
   &= 2[3(n-1)] - 3(n-2) \quad \text{since } f_n = 3n \\
   &= 6n - 6 - 3n + 6 = 3n.
   \end{aligned}$$

   $\therefore$ $\{f_n\}$, where $f_n = 3n$, is a solution of the recurrence relation.

2. Show that the sequence $\{f_n\}$ is a solution of the recurrence relation

   $f_n = -3f_{n-1} + 4f_{n-2}$ if $f_n = 2(-4)^n + 3$.

   Solution.

   $$\begin{aligned}
   f_n &= -3f_{n-1} + 4f_{n-2} \\
   &= -3\left[2(-4)^{n-1} + 3\right] + 4\left[2(-4)^{n-2} + 3\right] \\
   &= -6(-4)^{n-1} - 9 + 8(-4)^{n-2} + 12 \\
   &= -6(-4)^{n-1} + 8(-4)^{n-2} + 3 \\
   &= -6(-4)^{n-1} - 2(-4)^{n-1} + 3 \\
   &= 2(-4)^n + 3.
   \end{aligned}$$

   $\therefore$ $f_n = 2(-4)^n + 3$ is a solution of the recurrence relation.

Now, we discuss about a class of recurrence relations known as linear recurrence relations with constant coefficients.

2.5.5   Linear Recurrence Relation

A recurrence relation of the form

$$a_0 f_n + a_1 f_{n-1} + a_2 f_{n-2} + \cdots + a_k f_{n-k} = f(n) \tag{2.4}$$

where a_i's are constants, is called a linear recurrence relation with constant coefficients. The recurrence relation (2.4) is known as a kth-order recurrence relation, provided both a_0 and a_k are non-zero.

Note: The phrase "k^{th}-order" means that each term in the sequence depends only on the k previous terms.

Example 1:
Consider the Fibonacci sequence defined by the recurrence relation $f_n = f_{n-1} + f_{n-2}, n \geq 2$ and the initial conditions $f_0 = 0$ and $f_1 = 1$. The recurrence relation is called a second-order relation because f_n depends on the two previous terms of f_n.

Example 2:
Consider the recurrence relation $f(k) - 5f(k-1) + 6f(k-2) = 4k + 10$ defined for $k \geq 2$, together with the initial conditions $f(0) = \frac{7}{3}$ and $f(1) = 5$. Clearly, it is a second-order linear recurrence relation.

2.5.6   Homogenous Recurrence Relation

A kth-order linear relation is a homogenous recurrence relation if $f(n) = 0$ for all n. Otherwise, it is called non-homogenous.

Example 1:
Consider the recurrence relation $C(k) - 5C(k-1) + 8C(k-2) = 0$ together with the initial conditions $C(0) = 5$ and $C(1) = 2$. It is a second-order homogenous recurrence relation.

Example 2:
Which of the following recurrence relations are homogenous and which of them are non-homogenous?

$\quad$ (i) $f_n = f_{n-2}$.

$\quad$ (ii) $a_n = a_{n-1} + a_{n-3}$.

$\quad$ (iii) $b_n = b_{n-1} + 2$.

$\quad$ (iv) $s(n) = s(n-2) + s(n-4)$.

Solution.
The relations $f_n = f_{n-2}$, $a_n = a_{n-1} + a_{n-3}$, $s(n) = s(n-2) + s(n-4)$ are all homogenous, and the relation $b_n = b_{n-1} +$ is non-homogenous.

2.5.7 Recurrence Relations obtained from Solutions

Before giving an algorithm for solving a recurrence relation, we will examine a few recurrence relations that arise from certain closed form expressions. The procedure is illustrated by the following examples.

1. Form the recurrence relation given $f_n = 3 \cdot 5^n, n \geq 0$.

   Solution.
   If $n \geq 1$, then

   $$f_n = 3 \cdot 5^n = 3 \cdot 5 \cdot 5^{n-1}$$
   $$= 5 \cdot 3 \cdot 5^{n-1}$$
   $$= 5 f_{n-1}.$$

   $\therefore$ The recurrence relation is $f_n = 5 f_{n-1}$ with $f_0 = 3$.

2. Find the recurrence relation satisfying $y_n = A(3)^n + B(-2)^n$.

   Solution.
   Given $\quad y_n = A(3)^n + B(-2)^n$.

   $$\therefore \quad y_{n+1} = A(3)^{n+1} + B(-2)^{n+1} = 3A(3)^n - 2B(-2)^n$$
   $$y_{n+2} = A(3)^{n+2} + b(-2)^{n+2} = 9A(3)^n + 4B(-2)^n$$

   Eliminating A and B from the above equations,

   $$\begin{vmatrix} y_n & 1 & 1 \\ y_{n+1} & 3 & -2 \\ y_{n+2} & 9 & 4 \end{vmatrix} = 0$$

   Expanding along column 1,

   $$y_n(12 + 18) - y_{n+1}(4 - 9) + y_{n+2}(-2 - 3) = 0$$
   or $\quad 30y_n + 5y_{n+1} - 5y_{n+2} = 0$
   or $\quad 6y_n + y_{n+1} - y_{n+2} = 0$
   or $\quad y_{n+2} - y_{n+1} - 6y_n = 0$

   which is the required recurrence relation.

3. Find the recurrence relation satisfying $y_n = A(3)^n + B(-4)^n$.

   Solution.
   Given $\quad y_n = A(3)^n + B(-4)^n$.

   $$\therefore \quad y_{n+1} = 3A(3)^n - 4B(-4)^n$$
   $$y_{n+2} = 9A(3)^n + 16B(-4)^n$$

   Eliminating A and B from the above equations,

   $$\begin{vmatrix} y_n & 1 & 1 \\ y_{n+1} & 3 & -4 \\ y_{n+2} & 9 & 16 \end{vmatrix} = 0$$

Expanding along column 1,

$$y_n(48 + 36) - y_{n+1}(16 - 9) + y_{n+2}(-4 - 3) = 0$$

or $\quad 84y_n - 7y_{n+1} - 7y_{n+2} = 0$

or $\quad 12y_n - y_{n+1} - y_{n+2} = 0$

or $\quad y_{n+2} + y_{n+1} - 12y_n = 0$

which is the required recurrence relation.

4. Find the recurrence relation satisfying $y_n = (A + Bn)4^n$.

Solution.

Given $\quad y_n = (A + Bn)4^n = A(4)^n + Bn(4)^n$.

$$\therefore \quad y_{n+1} = 4A(4)^n + 4B(n + 1)(4)^n$$
$$y_{n+2} = 16A(4)^n + 16B(n + 2)(4)^n$$

Eliminating A and B from the above equations,

$$\begin{vmatrix} y_n & 1 & n \\ y_{n+1} & 4 & 4(n + 1) \\ y_{n+2} & 16 & 16(n + 2) \end{vmatrix} = 0$$

Expanding along column 1,

$$y_n[64(n + 2) - 64(n + 1)] - y_{n+1}[16n + 32 - 16n]$$
$$+ y_{n+2}[4n + 4 - 4n] = 0$$

or $\quad 64y_n - 32y_{n+1} + 4y_{n+2} = 0$

or $\quad y_{n+2} - 8y_{n+1} + 16y_n = 0$

which is the required recurrence relation.

2.6   Solving Linear Homogenous Recurrence Relations

Consider a linear homogenous recurrence relation of degree k with constant coefficients

$$f_n = a_1 f_{n-1} + a_2 f_{n-2} + \cdots + a_k f_{n-k}$$

where $a_1, a_2, \ldots, a_k$ are real numbers and $a_k \neq 0$. The basic approach for solving linear homogenous recurrence relations is to look for solutions of the form $f_n = r^n$, where r is a constant. Note that $f_n = r^n$ is a solution of the recurrence relation $f_n = a_1 f_{n-1} + a_2 f_{n-2} + \cdots + a_k f_{n-k}$ if and only if

$$r^n = c_1 r^{n-1} + c_2 r^{n-2} + \cdots + c_k r^{n-k}.$$

When both sides of this equation are divided by r^{n-k} and the right-hand side is subtracted from the left, we obtain

$$r^k - c_1 r^{k-1} - c_2 r^{k-2} - \cdots - c_{k-1} r - c_k = 0.$$

Consequently, the sequence $\{f_n\}$ with $f_n = r^n$ is a solution if and only if r is a solution of this last equation.

2.6.1 Characteristic Equation

The characteristic equation of the homogenous k^{th}-order linear recurrence relation $f_n + a_1 f_{n-1} + a_2 f_{n-2} + \cdots + a_k f_{n-k} = 0$ is the kth-degree polynomial equation

$$r^k + a_1 r^{k-1} + a_2 r^{k-2} + \cdots + a_{k-1} r^{k-1} + a_k = 0.$$

The solutions of this equation are called the characteristic roots of the recurrence relation.

Examples:

1. The characteristic equation of
   $Q(k) + 2Q(k-1) - 3Q(k-2) - 6Q(\text{k-4}) = 0$ is $r^4 + 2r^3 - 3r^2 - 6 = 0$.

   Note that the absence of $Q(k-3)$ term implies that there is no $r^{4-3} = r$ term in the characteristic equation.

2. The characteristic equation of $T(k) - 7T(k-2) + 6T(k-3) = 0$ is $r^3 - 7r + 6 = 0$, i.e. $r^3 - 7r + 6 = 0$, and $1, 2$, and -3 are the characteristic roots.

2.6.2 Algorithm for Solving k^{th}-order Homogenous Linear Recurrence Relations

Step 1:
If $f_n + a_1 f_{n-1} + a_2 f_{n-2} + \cdots + a_k f_{n-k} = 0$ is a given recurrence relation, then write its characteristic equation as $r^k + a_1 r^{k-1} + a_2 r^{k-2} + \cdots + a_{k-1} r + a_k = 0$.
Step 2:
Find all the characteristic roots of this equation.
Step 3:
Case (i): If there are k distinct roots, say $c_1, c_2, \ldots, c_k$, then the general solution of the recurrence relation is $f_n = A_1 c_1^k + A_2 c_2^k + \cdots + A_k C_k^k$.
Case (ii): Suppose that c_1 is a root of multiplicity m. Then, the corresponding solution is

$$f_n \left(A_1 r^{m-1} + A_2 r^{m-2} + \cdots + A_{m-2} r^2 + A_{m-1} r + A_m \right) c_1^r.$$

Step 4:
Use the boundary conditions to determine the constants $A_1, A_2, \ldots, A_k$.

2.6.3 Solved Problems

1. Solve the Fibonacci sequence $\{f_n\}$ defined by $f_n = f_{n-1} + f_{n-2}$ for $n \geq 2$ with the initial conditions $f_0 = 0$ and $f_1 = 1$.

   Solution.

   The characteristic equation of the given recurrence relation is

   $$r^2 - r - 1 = 0.$$

   Solving this equation, we get

   $$r = \frac{1 \pm \sqrt{1+4}}{2} = \frac{1 \pm \sqrt{5}}{2}.$$

   $$\therefore \qquad c_1 = \frac{1 + \sqrt{5}}{2}, \qquad c_2 = \frac{1 - \sqrt{5}}{2}.$$

   The general solution is

   $$f_n = A_1 c_1^n + A_2 c_2^n$$

   where A_1 and A_2 are constants.

   Given:

   $$f_0 = 1 \Rightarrow A_1 + A_2 = 0. \tag{2.5}$$

   $$f_1 = 1 \Rightarrow A_1 c_1 + A_2 c_2 = 1$$

   $$\Rightarrow A_1 \left(\frac{1 + \sqrt{5}}{2} \right) + A_2 \left(\frac{1 - \sqrt{5}}{2} \right) = 1. \tag{2.6}$$

   Solving (2.5) and (2.6), we get

   $$A_1 = \frac{1}{\sqrt{5}} \quad \text{and} \quad A_2 = -\frac{1}{\sqrt{5}}.$$

   $\therefore$ The solution is

   $$f_n = \frac{1}{\sqrt{5}} \left[\left(\frac{1 + \sqrt{5}}{2} \right)^n - \left(\frac{1 - \sqrt{5}}{2} \right)^n \right].$$

2. If the recurrence relation is $u_{n+1} - 2u_n = 0$, find the closed form expression (solution) for u_n.

   Solution.

   The characteristic equation is

   $$r - 2 = 0$$

   $$\Rightarrow r = 2.$$

   The general solution is

   $$u_n = A \cdot 2^n$$

   where A is a constant.

3. Find $f(n)$ if $f(n) = 7f(n-1) - 10f(n-2)$, given that $f(0) = 4$ and $f(1) = 17$.

Solution.

The characteristic equation is

$$r^2 - 7r + 10 = 0$$

$$\Rightarrow r = 2, 5$$

$$\therefore \quad c_1 = 2, c_2 = 5.$$

The general solution is

$$f(n) = A_1 c_1^n + A_2 c_2^n$$
$$= A_1 2^n + A_2 5^n.$$

Given:

$$f(0) = 4 \Rightarrow A_1 + A_2 = 4. \tag{2.7}$$

$$f(1) = 17 \Rightarrow 2A_1 + 5A_2 = 17. \tag{2.8}$$

Solving (2.7) and (2.8), we get

$$A_1 = 3 \quad \text{and} \quad A_2 = 3.$$

$$\therefore \quad f(n) = 2^n + 3(5)^n.$$

4. Find $T(k)$ if $T(k) - 7T(k-2) + 6T(k-3) = 0$, where $T(0) = 8$, $T(1) = 6$, and $T(2) = 22$.

Solution.

The characteristic equation is

$$r^3 - 7r + 6 = 0.$$

$$
\begin{array}{c|cccc}
1 & 1 & 0 & -7 & 6 \\
  & 0 & 1 & 1 & 6 \\
\hline
2 & 1 & 1 & -6 & \boxed{0} \\
  & 0 & 2 & 6 & \\
\hline
  & 1 & 3 & \boxed{0} &
\end{array}
$$

The characteristic roots are

$$c_1 = 1, \quad c_2 = 2, \quad c_3 = -3.$$

The general solution is

$$T(k) = A_1 c_1^k + A_2 c_2^k + A_3 c_3^k$$
$$T(k) = A_1 + A_2 (2)^k + A_3 (-3)^k.$$

Given:

$$T(0) = 8 \Rightarrow A_1 + A_2 + A_3 = 8. \tag{2.9}$$

$$T(1) = 6 \Rightarrow A_1 + 2A_2 - 3A_3 = 6. \tag{2.10}$$

$$T(2) = 22 \Rightarrow A_1 + 4A_2 + 9A_3 = 22. \tag{2.11}$$

Solving (2.9), (2.10) and (2.11), we get

$$A_1 = 5, \quad A_2 = 2, \quad A_3 = 1.$$

$$\therefore \qquad T(k) = 5 + 2(2)^k + 1(-3)^k$$

$$\text{or} \qquad T(k) = 5 + 2^{k+1} + (-3)^k.$$

5. Solve $f_k - 8f_{k-1} + 16f_{k-2} = 0$ where $f_2 = 16$ and $f_3 = 80$.

   Solution.

   The characteristic equation is

   $$r^2 - 84 + 16 = 0$$
   $$\Rightarrow r = 4, 4 \quad \text{(repeated)}.$$

   The general solution is

   $$f_k = (A_1 + A_2 k)4^k.$$

Given:

$$f_2 = 16 \Rightarrow (A_1 + 2A_2)16 = 16$$
$$\Rightarrow A_1 + 2A_2 = 1. \tag{2.12}$$

$$f_3 = 80 \Rightarrow (A_1 + 3A_2)64 = 80$$
$$\Rightarrow 4(A_1 + 3A_2) = 5$$
$$\Rightarrow 4A_1 + 12A_2 = 5. \tag{2.13}$$

Solving (2.12) and (2.13), we get

$$A_1 = \frac{1}{2}, \qquad A_2 = \frac{1}{4}.$$

$\therefore$ The solution is

$$f_k = \left(\frac{1}{2} + \frac{1}{4}k\right)4^k = (2 + k)4^{k-1}.$$

6. Find a solution to the recurrence relation
   $C_n = -3C_{n-1} - 3C_{n-2} - C_{n-3}$ for $n \geq 3$ with initial conditions
   $C_0 = 1, C_1 = -2,$ and $C_2 = 1$.

Solution.
The characteristic equation is

$$r^3 + 3r^2 + 3r + 1 = 0$$
$$\Rightarrow (r+1)^3 = 0.$$

$\therefore \quad r = -1$ is a characteristic root of multiplicity 3.
The general solution is

$$C_n = \left(A_1 + A_2 n + A_3 n^2\right)(-1)^n.$$

Given:

$$C_0 = 1 \Rightarrow A_1 = 1. \tag{2.14}$$
$$C_1 = -2 \Rightarrow -(A_1 + A_2 + A_3) = -2. \tag{2.15}$$
$$C_2 = 1 \Rightarrow A_1 + 2A_2 + 4A_3 = 1. \tag{2.16}$$

Solving (2.14), (2.15), and (2.16), we get

$$A_1 = 1, \quad A_2 = 2, \quad A_3 = -1.$$

$\therefore \quad$ The solution is

$$C_n = \left(1 + 2n - n^2\right)(-1)^n.$$

2.7 Solving Linear Non-homogenous Recurrence Relations

The solution of a linear non-homogenous recurrence relation with constant coefficients is the sum of the two parts, the homogenous solution, which satisfies the recurrence relation when the right-hand side of the equation is set to 0, and the particular solution, which satisfies the difference equation with $f(n)$ on the right-hand side.

There is no general procedure for determining the particular solution of a difference equation. However, in simple cases, this solution can be obtained by the method of inspection. To determine the particular solution, we use the following rules:

Rule 1:
When $f(n)$ is of the form of a polynomial of degree m in n,

$$k_0 + k_1 n + k_2 n^2 + k_3 n^3 + \cdots + k_{m-1} n^{m-1} + k_m n^m,$$

the corresponding particular solution will be of the form

$$Q_0 + Q_1 n + Q_2 n^2 + Q_3 n^3 + \cdots + Q_{m-1} n^{m-1} + Q_m n^m.$$

Rule 2:

When $f(n)$ is of the form

$$\left(k_0 + k_1 n + k_2 n^2 + \cdots + k_{m-1} n^{m-1} + k_m n^m\right) a^n,$$

the corresponding particular solution is of the form

$$\left(Q_0 + Q_1 n + Q_2 n^2 + \cdots + Q_{m-1} n^{m-1} + Q_m n^m\right) a^n$$

if a is not a characteristic root of the recurrence relation.

Rule 3:

If a is a characteristic root of multiplicity $r - 1$, when $f(n)$ is of the form

$$\left(k_0 + k_1 n + k_2 n^2 + \cdots + K_{m-1} n^{m-1} + K - mn^m\right) a^n,$$

the corresponding particular solution is of the form

$$r^{n-1}\left(Q_0 + Q_1 n + Q_2 n^2 + \cdots + Q_{m-1} n^{m-1} + Q_m n^m\right) a^n.$$

Note:

The general solution of the recurrence relation is the sum of the homogenous solution and particular solution. If no initial conditions are given, then you have finished. If m initial conditions are given, obtain m linear equations in m unknowns and solve the system, if possible, to get a complete solution.

2.7.1   Solved Problems

1. Solve $S(k) - S(k-1) - 6S(k-2) = -30$ where $S(0) = 20$, $S(1) = -5$.

   Solution.

   The associated homogenous relation is

   $$S(k) - S(k-1) - 6S(k-2) = 0.$$

   The characteristic equation is

   $$r^2 - r - 6 = 0.$$

   The characteristic roots are $r = -2, 3$.

   The homogenous solution is

   $$A_1(-2)^k + A_2(3)^k.$$

   Since the right-hand side of

   $$S(k) - S(k-1) - 6S(k-2) = -30 \qquad (2.17)$$

is a constant, by Rule 1, the particular solution will be a constant, say Q. Substituting Q into (2.17), we obtain

$$Q - Q - 6Q = -30$$
$$\Rightarrow Q = 5.$$

∴   The general solution is

$$S(k) = A_1(-2)^k + A_2(3)^k + 5.$$

Using the initial conditions, we have

$$S(0) = 20 \Rightarrow A_1 + A_2 + 5 = 20. \tag{2.18}$$

$$S(1) = -5 \Rightarrow = 2A_1 + 3A_2 + 5 = -5. \tag{2.19}$$

Solving (2.18) and (2.19), we get $A_1 = 11$, $A_2 = 4$.

∴   The complete solution is

$$S(k) = 11(-2)^k + 4(3)^k + 5.$$

2. Solve the recurrence relation $f_n - 5f_{n-1} + 6f_{n-2} = 1$.

Solution.

The associated homogenous relation is

$$f_n - 5f_{n-1} + 6f_{n-2} = 0.$$

The characteristic equation is $r^2 - 5r + 6 = 0$.

The characteristic roots are $r = 2, 3$.

The homogenous solution is $A_2(2)^n + A_2(3)^n$.

Since the right-hand side of the given relation is 1 (a constant), by Rule 1, the particular solution will also be a constant, say Q.

$$Q - 5Q + 6Q = 1$$
$$\Rightarrow \qquad Q = \frac{1}{2}.$$

∴   The complete solution is

$$f_n = A_1(2)^n + A_2(3)^n + \frac{1}{2}.$$

3. Find the particular solution of the recurrence relation

$$f(n) + 5f(n-1) + 6f(n-2) = 3n^2 - 2n + 1. \tag{2.20}$$

Solution.

By rule 2, the particular solution is of the form

$$Q_0 + Q_1 n + Q_2 n^2. \qquad (2.21)$$

Substituting (2.21) in (2.20), we get

$$\left(Q_0 + Q_1 n + Q_2 n^2\right) + 5\left[Q_0 + Q_1(n-1) + Q_2(n-1)^2\right]$$
$$+ 6\left[Q_0 + Q_1(n-2) + Q_2(n-2)^2\right] = 3n^2 - 2n + 1$$

which simplifies to

$$(12Q_0 - 17A_1 + 29Q_2) + (12Q_1 - 34Q_2)n + 12Q_2 n^2$$
$$= 3n^2 - 2n + 1. \qquad (2.22)$$

Comparing both sides of (2.22), we obtain

$$12Q_2 = 3; \quad 12Q_1 - 34Q_2 = -2; \quad 12Q_0 - 17Q_1 + 29Q_2 = 1$$

which gives

$$Q_2 = \frac{1}{4}; \quad Q_1 = \frac{13}{24}; \quad Q_0 = \frac{71}{288}.$$

$\therefore$ The particular solution is

$$\frac{71}{288} + \frac{13}{24}n + \frac{1}{4}n^2.$$

4. Solve $a_r + 5a_{r-1} = 9$ with initial condition $a_0 = 6$.

Solution.

The associated homogenous relation is

$$a_r + 5a_{r-1} = 0.$$

The characteristic equation is $r + 5 = 0$.

The characteristic root is $r = -5$.

The homogenous solution is $A(-5)^r$.

Since the right-hand side of the given relation is a constant, the particular solution will also be a constant Q. Substituting in the relation, we get

$$Q + 5Q = 9$$
$$\Rightarrow \qquad Q = \frac{3}{2}.$$

$\therefore$ The general solution is

$$A_r = A(-5)^r + \frac{3}{2}.$$

Given:

$$a_0 = 6 \Rightarrow A + \frac{3}{2} = 6$$

$$\Rightarrow A = \frac{9}{2}.$$

∴ The complete solution is

$$a_r = \frac{9}{2}(-5)^r + \frac{3}{2}.$$

5. Solve the recurrence relation $f(n) - 7f(n-1) + 10f(n-2) = 6 + 8n$ with $f(0) = 1$ and $f(1) = 2$.

Solution.

The characteristic equation is

$$r^2 - 7r + 10 = 10$$

$$\Rightarrow \quad r = 2, 5.$$

Homogenous solution is $A_1(2)^n + A_2(5)^n$.

By Rule 1, the particular solution is of the form $Q_0 + Q_1^n$. Substituting in the given relation, we obtain

$$(Q_0 + Q_1^n) - 7[Q_0 + Q_1(n-1)] + 10[Q_0 + Q_1(n-2)] = 6 + 8n.$$

Comparing both sides, we obtain

$$4Q_0 - 13Q_1 = 6 \quad \text{and} \quad 4Q_1 = 8$$

which yield $Q_0 = 8$ and $Q_1 = 2$.

∴ The particular solution is $8 + 2n$.

The general solution is

$$f(n) - A_1(2)^n + A_2(5)^n + 8 + 2n.$$

Given:

$$f(0) = 1 \Rightarrow A_1 + A_2 + 8 = 1. \tag{2.23}$$

$$f(1) = 2 \Rightarrow 2A_1 + 5A_2 + 10 = 2. \tag{2.24}$$

Solving (2.23) and (2.24), we get $A_1 = -9, \quad A_2 = 2$.

∴ The complete solution is

$$f(n) = -9(2)^n + 2(5)^n + 8 + 2n.$$

6. Find the particular solution of the recurrence relation
   $a_n + 5a_{n-1} + 6a_{n-2} = 42(4)^n$.

   Solution.
   The characteristic equation is $r^2 + 5r + 6 = 0$.

   The characteristic roots are $r = -2, -3$.

   Since 4 is not a characteristic root, by Rule 2, we assume that the general form of the particular solution is $Q \cdot (4)^n$. Substituting in the given relation, we obtain

   $$Q \cdot (4)^n + 5Q \cdot (4)^{n-1} + 6Q \cdot (4)^{n-2} = 42(4)^n$$
   $$\Rightarrow \quad Q \cdot 4^{n-2}[16 + 20 + 6] = 42(4)^n$$
   $$\Rightarrow \quad Q \cdot 4^{n-2}(42) = 42(4)^n$$
   $$\Rightarrow \quad Q = 16.$$

   $\therefore$ The particular solution is $16(4)^n = 4^{n+2}$.

7. Find the particular solution of the recurrence relation
   $f_n + f_{n-1} = 3n2^n$.

   Solution.
   The characteristic equation is $r + 1 = 0$.

   The characteristic root is $r = -1$.

   Since 2 is not a characteristic root, by Rule 2, the general form of the particular solution is $(Q_0 + Q_1 n)2^n$.

   Substituting in the given relation, we obtain

   $$(Q_0 + Q_1 n)2^n + [Q_0 + Q_1(n-1)]2n - 1 = 3n2^n$$

   which simplifies to

   $$Q_0 2^n + Q_1 n 2^n + \frac{1}{2}Q_0 2^n + \frac{1}{2}Q_1 n 2^n - \frac{1}{2}Q_1 n 2^n = 3n2^n$$
   $$\Rightarrow \quad \left(\frac{3}{2}Q_0 - \frac{1}{2}Q_1\right)2^n + \frac{3}{2}Q_1 n 2^n = 3n2^n$$
   $$\Rightarrow \quad \frac{3}{2}Q_0 - \frac{1}{2}Q_1 = 0$$
   and $\qquad \frac{3}{2}Q_1 = 3.$

   Solving, we get $Q_0 = \frac{2}{3}$ and $Q_1 = 2$.

   $\therefore$ The particular solution is $\left(\dfrac{2}{3} + 2n\right)2^n$.

8. Find the particular solution of the recurrence relation
   $f(n) - 2f(n-1) = 3 \cdot 2^n$.

Solution.

The characteristic equation is $r - 2 = 0$.

The characteristic equation is $r = 2$.

Since $r = 2$ is the characteristic root of multiplicity 1, by Rule 3, the general form of the particular solution is $Qn \cdot 2^n$.

Substituting in the given relation, we obtain

$$Qn2^n - 2\left[Q \cdot (n-1)2^{n-1}\right] = 3 \cdot 2^n$$
$$\Rightarrow \quad Qn2^n - Qn2^n + Q2^n = 3 \cdot 2^n$$
$$\Rightarrow \quad Q = 3.$$

$\therefore$ The particular solution is $3n(2^n)$.

9. Find the general solution of

$$f(n) - 3f(n-1) - 4f(n-2) = 4^n. \tag{2.25}$$

Solution.

The associated homogenous relation is

$$f(n) - 3f(n-1) - 4f(n-2) = 0.$$

The characteristic equation is $r^2 - 3r - 4 = 0$.

The characteristic roots are $r = -1, 4$.

The homogenous solution is $A_1(-1)^n + A_2(4)^n$. Since 4 is a characteristic root, by Rule 3, we assume that the general form of the particular solution is $Qn4^n$.

Substituting in (2.25), we obtain

$$Qn4^n - 3Q \cdot (n-1)4^{n-1} - 4Q \cdot (n-2)4^{n-2} = 4^n$$
$$\Rightarrow \quad Qn4^n - 3Qn4^{n-1} + 3Q4^{n-1} - 4Qn4^{n-2} + 8Q4^{n-2} = 4^n$$
$$\Rightarrow \quad (16Qn - 12Qn + 12Q - 4Qn + 8Q)4^{n-2} = (16)4^{n-2}$$
$$\Rightarrow \quad 20Q = 16$$
$$\Rightarrow \quad Q = \frac{4}{5}.$$

$\therefore$ The particular solution is $\frac{4}{5}n(4)^n$.

The general solution of the given recurrence relation is

$$f(n) = A_1(-1)^n + A_2(4)^n + \frac{4}{5}n4^n.$$

Remark:

What if the characteristic equation gives rise to complex roots? Here, our methods are still valid, but the method for expressing the solutions of the recurrence relations is different. Since an understanding of these representations require some background in complex numbers, we suggest that an interested reader refer to a more advanced treatment of recurrence relations.

2.7.2   Problems for Practice

1. Find the general solution of the following recurrence relations.

   (i) $f_n - 3f_{n-1} - 10f_{n-2} = 0$

   (ii) $f_{n+2} + 6f_{n+1} + 9f_n = 0$

   (iii) $2f_n + 2f_{n-1} - f_{n-2} = 0$

   (iv) $f_n - 3f_{n-1} - 4f_{n-2}$.

2. Solve the following recurrence relations.

   (i) $f(n) - 10f(n-1) + 9f(n-2) = 0; f(0) = 3; f(1) = 11$

   (ii) $f(n) - 9f(n-1) + 18f(n-2) = 0; f(0) = 1; f(1) = 4$

   (iii) $f(n+2) - 8f(n+1) + 16f(n) = 0; f(0) = 0; f(1) = 8$

   (iv) $f(n) - 3f(n-1) + 3f(n-2) - f(n-3) = 0; f(1) = 0;$
   $f(2) = 1; f(3) = 0$

   (v) $f(n+2) - 2f(n+1) + f(n) = 0; f(0) = 1; f(1) = 2$

   (vi) $f(n) - 20f(n-1) + 100f(n-2) = 0; f(0) = 2; f(1) = 30$.

3. Find the recurrence relation satisfying

   (i) $y_n = A(3)^n + B(8)^n$

   (ii) $y_n = (A + Bn)(-2)^n$

   (iii) $y_n = (A + Bn)(6)^n$

   (iv) $y_n = A(3)^n + B(5)^n$

   (v) $y_n = 2(3)^n$.

4. Solve the following set of recurrence relations with the initial conditions.

   (i) $y_n - 2y_{n-1} = 6n; y_1 = 2$

   (ii) $y_{n+2} + 2y_{n+1} - 15y_n = 6n + 10; y_0 = 1; y_1 = -\frac{1}{2}$

   (iii) $y_{n+1} + 2y_n = 3 + 4^n; y_0 = 2$

   (iv) $y_{n+2} - 2y_{n+1} + y_n = 1; y_0 = 1; y_1 = \frac{1}{2}$

   (v) $y_{n+1} + y_n = 5; y_0 = 1$

   (vi) $y_n - 3y_{n-1} + 2y_{n-2} - n^2; y_0 = 0; y_1 = 0$

   (vii) $y_n - 4y_{n-1} + 4y_{n-2} = 3n + 2^n; y_0 = 1; y_1 = 1$

   (viii) $y_n - 5y_{n-1} = 5^n; y_0 = 3$.

2.8   Generating Functions

In this section, we will show how recurrence relations can be solved using the powerful generating function method. Generating function is an important tool in discrete mathematics, and its use is by no means confined to the solution of recurrence relations.

If $a_0, a_1, a_2, \ldots, a_n$ is a finite sequence of numbers, the generating function for the a_n's is the polynomial

$$G(z) = \sum_{k=0}^{n} a_k z^k = a_0 + a_1 z + a_2 z^2 + \cdots + a_n z^n$$

where z is an indeterminate (that is, an abstract) symbol. If $a_0, a_1, a_2, \ldots, a_n, \ldots$ is an infinite sequence of numbers, its generating function is defined to be

$$G(z) = \sum_{k=0}^{\infty} a_k z^k = a_0 + a_1 z + a_2 z^2 + \cdots.$$

The symbol z is just the name given to a variable and has no special significance. For any sequence $\{a_n\}$, we write $G(z)$ to denote the generating function of $\{a_n\}$. Clearly, given a sequence, we can easily obtain its generating function and its converse. For example, the generating function of $a_n = \alpha^n$, $n \geq 0$ is

$$\alpha^0 + \alpha z + \alpha^2 z^2 + \alpha^3 z^3 + \cdots \qquad (2.26)$$

We note that the infinite series (2.26) can be written in closed form as $\dfrac{1}{1 - \alpha z}$ which is a rather compact way to represent the sequence $\{a_n\}$ or $(a, \alpha, \alpha^2, \ldots)$.

2.8.1   Solved Problems

1. Find the generating function for the sequence $1, 1, 1, 1, 1, 1$.

    Solution.
    By definition, the generating function of $1, 1, 1, 1, 1, 1$ is

    $$G(z) = 1 + z + z^2 + z^3 + z^4 + z^5 = \frac{z^6 - 1}{z - 1}.$$

2. Let n be a positive integer. Let $a_k = C(n, k)$ for $k = 0, 1, 2, \ldots, n$. Find the generating function for the sequence $a_0, a_1, \ldots, a_n$.

    Solution.
    The generating function for this sequence is

    $$G(z) = C(n, 0) + C(n, 1)z + C(n, 2)z^2 + \cdots + c(n, n)z^n$$
    $$= (1 + z)^n, \quad \text{by binomial theorem.}$$

3. Find the generating function for the infinite sequence $1, \alpha, \alpha^2, \alpha^3, \ldots,$ where α is a fixed constant.

Solution.

The generating function for this sequence is

$$G(z) = 1 + \alpha z + \alpha^2 z^2 + \alpha^3 z^3 + \ldots$$

$$= \frac{1}{1 - \alpha z}.$$

4. Find the generating function for $f_n = 3^n$, $n \geq 0$ in closed form.

Solution.

The generating function for this sequence is

$$G(z) = 1 + 3z + (3z)^2 + (3z)^3 + \ldots$$

$$= \frac{1}{1 - 3z}.$$

5. Find the generating function (in closed form) of the Fibonacci sequence $\{f_n\}$ defined by

$$f_n = f_{n-1} + f_{n-2}; f_0 = 0; f_1 = 1.$$

Solution.

The generating function is

$$G(z) = f_0 + f_1 z + f_2 z^2 + f_3 z^3 + \cdots = \sum_{n=0}^{\infty} f_n z^n. \qquad (2.27)$$

Consider $f_n = f_{n-1} + f_{n-2}; n \geq 0.$
Multiplying both sides by z^n and summing over all $n \geq 2$, we get

$$\sum_{n=2}^{\infty} f_n z^n = \sum_{n=2}^{\infty} f_{n-1} z^n + \sum_{n=2}^{\infty} f_{n-2} z^n. \qquad (2.28)$$

Consider the first sum

$$\sum_{n=2}^{\infty} f_n z^n = f_2 z^2 + f_3 z^3 + \ldots$$

$$= G(z) - f_0 - f_1 z \quad [\text{using } (2.28)].$$

Similarly,

$$\sum_{n=2}^{\infty} f_{n-1} z^n = f_1 z^2 + f_2 z^3 + \ldots$$

$$= z(f_1 z + f_2 z^2 + \ldots)$$
$$= z[G(z) - f_0] \quad [\text{using } (2.28)]$$

and

$$\sum_{n=2}^{\infty} f_{n-2} z^n = f_0 z^2 + f_1 z^3 + f_2 z^4 + \ldots$$

$$= z^2 (f_0 + f_1 z + f_2 z^2 + \ldots)$$
$$= z^2 [G(z)] \quad [\text{using (2.28)}].$$

Substituting these expressions in (2.28), we obtain

$$G(z) - f_0 - f_1 z = z[G(z) - f_0] + z^2 G(z).$$

Since $f_0 = 0$ and $f_1 = 1$, we get

$$G(z) - z = zG(z) + z^2 G(z)$$
$$\Rightarrow \quad G(z)(1 - z - z^2) = z$$
$$\Rightarrow \quad G(z) = \frac{z}{1 - z - z^2}$$

which is the required generating function.

6. Find the generating function of the sequence $y_0, y_1, \ldots, y_n$ defined as follows:

$$y_n + 2y_{n-1} - 15y_{n-2} = 0 \quad \text{for} \quad n \geq 2 \quad \text{with} \quad y_0, y_1 = 1. \quad (2.29)$$

Solution.
The generating function is

$$G(z) = y_0 + y_1 z + y_2 z^2 + \cdots = \sum_{n=2}^{\infty} y_n z^n.$$

Multiplying (2.29) by z^n and summing over all $n \geq 2$, we get

$$\sum_{n=2}^{\infty} y_n z^n + 2 \sum_{n=2}^{\infty} y_{n-1} z^n - 15 \sum_{n=2}^{\infty} y_{n-2} z^n = 0$$
$$\Rightarrow \quad [G(z) - y_0 - y_1^2] + 2[z\{G(z) - y_0\}] - 15z^2 G(z) = 0.$$

Since $y_0 = 0, y_1 = 1$, we get

$$[G(z) - 1] + 2[zG(z) - 15z^2 G(z)] = 0$$
$$\Rightarrow \quad G(z)(1 + 2z - 15z^2) = z$$
$$\Rightarrow \quad G(z) = \frac{z}{1 + 2z - 15z^2}$$
$$\Rightarrow \quad G(z) = \frac{z}{(1 - 3z)(1 + 5z)}.$$

2.8.2 Solution of Recurrence Relations Using Generating Function

We can find the solution to a recurrence relation with its initial conditions by finding an explicit formula for the associated generating function. This is illustrated in the following examples. The following are some important fundamental results useful for solved examples presented below:

1. $(1+x)^n = \sum_{r=0}^{\infty} \frac{n(n-1)...(n-r+1)}{r!} x^r$

2. $(1+x)^{-n} = \sum_{r=0}^{\infty} (-1)^r \frac{n(n-1)...(n-r+1)}{r!}$

3. $(1+x)^{-1} = \sum_{r=0}^{\infty} (-1)^r x^r = 1 + x + x^2 + \ldots$

4. $(1+x)^{-2} = \sum_{r=0}^{\infty} (-1)^r (r+1) x^r = 1 - 2x + 3x^2 - 4x^3 + \ldots$

5. $(1-x)^{-1} = \sum_{r=0}^{\infty} x^r = 1 + x + x^2 + \ldots$

6. $(1-x)^{-2} = \sum_{r=0}^{\infty} (r+1) x^r = 1 + 2x + 3x^2 + 4x^3 + \ldots$

7. $e^x = \sum_{r=0}^{\infty} \frac{x^r}{r!} = 1 + \frac{x}{1!} + \frac{x^2}{2!} + \frac{x^3}{3!} + \ldots$

2.8.3 Solved Problems

1. Using the generating function, solve the recurrence relation $f_n = 3f_{n-1}$, for $n = 1, 2, 3 \ldots$ and initial condition $f_0 = 2$.

   Solution.
   Let the generating function be

$$G(z) = \sum_{n=0}^{\infty} f_n z^n. \qquad (2.30)$$

   Multiplying the given relation by z^n and summing for all $n \geq 1$, we obtain

$$\sum_{n=1}^{\infty} f_n z^n = 3 \sum_{n=1}^{\infty} f_{n-1} z^n. \qquad (2.31)$$

   The first sum

$$\sum_{n=1}^{\infty} f_n z^n = f_1 z + f_2 z^2 + \cdots = G(z) - f_0 \quad [\text{using } (2.30)]$$

   and the second sum

$$\sum_{n=1}^{\infty} f_{n-1} z^n = f_0 z + f_1 z^2 + f_2 z^3 + \ldots$$

$$= z[f_0 + f_1 z + f_2 z^2 + \ldots]$$
$$= zG(z) \quad [\text{using } (2.30)].$$

Hence, (2.31) becomes

$$G(z) - f_0 = 3zG(z)$$
$$\Rightarrow \quad G(z) - 2 = 3zG(Z)$$

$$\Rightarrow \quad G(z) = \frac{2}{1 - 3z} = 2(1 - 3z)^{-1} = 2\sum_{n=0}^{\infty}(3z)^n$$

$$= \sum_{n=0}^{\infty} 2 \cdot 3^n z^n. \tag{2.32}$$

Comparing (2.30) and (2.32), we get

$$f_n = 2 \cdot 3^n$$

which is the required solution.

2. Using generating function, solve the recurrence relation $y_n = 3y_{n-1} + 2$; $n \geq 1$ with $y_0 = 1$.

Solution.

Let the generating function be

$$G(z) = \sum_{n=0}^{\infty} y_n z^n. \tag{2.33}$$

Given:
$$y_n = 3y_{n-1} + 2. \tag{2.34}$$

Multiplying both sides of (2.34) by z^n and summing for $n \geq 1$, we get

$$\sum_{n=1}^{\infty} y_n z^n = 3\sum_{n=1}^{\infty} y_{n-1} z^n + 2\sum_{n=1}^{\infty} z^n. \tag{2.35}$$

Consider the first sum

$$\sum_{n=1}^{\infty} y_n z^n = y_1 z + y_2 z^2 + \cdots = G(z) - y_0 \quad [\text{using } (2.33)],$$

the second sum

$$\sum_{n=1}^{\infty} y_{n-1} z^n = y_0 z + y_1 z^2 + y_2 z^3 + \cdots$$

$$= z[y_0 + y_1 z + y_2 z^2 + \cdots]$$
$$= zG(z) \quad [\text{using } (2.33)],$$

and the third sum

$$\sum_{n=1}^{\infty} = z + z^2 + z^3 + \dots$$
$$= z(1 + z + z^2 + \dots)$$
$$= z(1 - z)^{-1}$$
$$= \frac{z}{1 - z}.$$

Hence, (2.35) becomes

$$G(z) - y_0 = 3zG(z) + \frac{2z}{1 - z}.$$

Using $y_0 = 1$,

$$G(z)(1 - 3z) = 1 + \frac{2z}{1 - z} = \frac{1 + z}{1 - z}.$$
$$\Rightarrow \quad G(z) = \frac{1 + z}{(1 - z)(1 - 3z)}.$$

Using partial fraction,

$$\frac{1 + z}{(1 - z)(1 - 3z)} = \frac{A}{1 - 3z} + \frac{B}{1 - z}$$
$$\Rightarrow \quad 1 + z = A(1 - z) + B(1 - 3z).$$

Put $z = 1 \Longrightarrow 2 = -2B \Longrightarrow B = -1$.
Put $z = \dfrac{1}{3} \Longrightarrow \dfrac{4}{3} = \dfrac{2}{3}A \Longrightarrow A = 2$.

$$\therefore \quad G(z) = \frac{2}{1 - 3z} - \frac{1}{1 - z} = 2(1 - 3z)^{-1} - (1 - z)^{-1}$$
$$= 2\sum_{n=0}^{\infty}(3z)^n - \sum_{n=0}^{\infty}z^n$$
$$= \sum_{n=0}^{\infty}(2 \cdot 3^n - 1)\, z^n. \qquad (2.36)$$

Comparing (2.33) and (2.36), we get

$$y_n = 2 \cdot 3^n - 1.$$

3. Using generating function, solve the difference equation

$$y_{n+2} - 4y_{n+1} + 3y_n = 0;\ y_0 = 2;\ y_1 = 4.$$

Solution.

Let the generating function be

$$G(z) = \sum_{n=0}^{\infty} y_n z^n. \tag{2.37}$$

Multiplying the given relation by z^n and summing for $n \geq 0$, we get

$$\sum_{n=0}^{\infty} y_{n+2} z^n - 4 \sum_{n=0}^{\infty} y_{n+1} z^n + 3 \sum_{n=0}^{\infty} y_n z^n = 0. \tag{2.38}$$

Consider the first sum

$$\sum_{n=2}^{\infty} y_{n+2} z^n = y_2 + y_3 z + y_4 z^2 + y_5 z^3 + \ldots$$

$$= \frac{1}{z^2} \left(y_2 z^2 + y_3 z^3 + y_4 z^4 + \ldots \right)$$

$$= \frac{1}{2} [G(z) - y_0] \quad [\text{using } (2.37)],$$

the second sum

$$\sum_{n=0}^{\infty} = y_1 + y_2 + y_3 z^2 + y_4 z^3 + \ldots$$

$$= \frac{1}{z} [y_1 z + y_2 z^2 + y_3 z^3 + \ldots]$$

$$= \frac{1}{z} [G(z) - y_0] \quad [\text{using } (2.37)],$$

and the third sum

$$\sum_{n=0}^{\infty} y_n z^n = G(z) \quad [\text{using } (2.37)].$$

∴ (2.38) becomes

$$\frac{1}{z^2} [G(z) - y_0 - y_1 z] - \frac{4}{z} [G(z) - y_0] + 3G(z) = 0.$$

Since $y_0 = 2, y_1 = 4$,

$$\frac{1}{z^2} [G(z) - 2 - 4z] - \frac{4}{z} [G(z) - 2] + 3G(z) = 0$$

$$\implies [G(z) - 4z - 2] - 4z[G(z) - 2] + 3z^2 G(z) = 0$$

$$\implies G(z)[1 - 4z + 3z^2] = 2 - 4z$$

$$\implies G(z) = \frac{2 - 4z}{1 - 4z + 3z^2} = \frac{2 - 4z}{(1 - z)(1 - 3z)}.$$

Using partial fractions,

$$\frac{2 - 4z}{(1 - z)(1 - 3z)} = \frac{A}{1 - z} + \frac{B}{1 - 3z}$$
$$\implies \quad 2 - 4z = A(1 - 3z) + B(1 - z).$$

Put $z = 1 \implies -2 = -2A \implies A = 1$.
Put $z = \frac{1}{3} \implies \frac{2}{3} = \frac{2}{3}B \implies B = 1$.

$$\therefore G(z) = \frac{1}{1 - z} + \frac{1}{1 - 3z} = (1 - z)^{-1} + (1 - 3z)^{-1}$$

$$= \sum_{n=0}^{\infty} z^n + \sum_{n=0}^{\infty} (3z)^n$$

$$= \sum_{n=0}^{\infty} (1 + 3^n) z^n. \qquad (2.39)$$

Comparing (2.37) and (2.39), the required solution is

$$y_n = 1 + 3^n.$$

4. Using generating function, solve the difference equation

$$y_{n+2} - 6y_{n+1} + 8y_n = 0, y_0 = 1, y_1 = 4.$$

Solution.
Let the generating function be

$$G(z) = \sum_{n=0}^{\infty} y_n z^n. \qquad (2.40)$$

Multiplying the given equation by z^n and summing for $n \geq 0$, we get

$$\sum_{n=0}^{\infty} y_{n+2} z^n - 6 \sum_{n=0}^{\infty} y_{n+1} z^n + 8 \sum_{n=0}^{\infty} y_n z^n = 0. \qquad (2.41)$$

Consider the first sum

$$\sum_{n=0}^{\infty} y_{n+2} z^n = y_2 + y_3 z + y_4 z^2 + y_5 z^3 + \dots$$

$$= \frac{1}{z^2} \left(y_2 z^2 + y_3 z^3 + y_4 z^4 + \dots \right)$$

$$= \frac{1}{z^2} [G(z) - y_0 - y_1 z] \quad [\text{using } (2.40)],$$

the second sum

$$\sum_{n=0}^{\infty} y_{n+1} z^n = y_1 + y_2 z + y_3 z^2 + \ldots$$

$$= \frac{1}{z}(y_1 z + y_2 z^2 + y_3 z^3 + \ldots)$$

$$= \frac{1}{z}[G(z) - y_0] \quad [\text{using } (2.40)],$$

and the third sum

$$\sum_{n=0}^{\infty} y_n z^n = G(z) \quad [\text{using } (2.40)].$$

Hence, (2.41) becomes

$$\frac{1}{z^2}[G(z) - y_0 - y_1 z] - \frac{6}{z}[G(z) - y_0] + 8G(z) = 0$$

$$\implies \quad [G(z) - y_0 - y_1 z] - 6z[G(z) - y_0] + 8z^2 G(z) = 0$$

$$\implies \quad G(z)[1 - 6z + 8z^2] = 1 - 2z \quad (\because y_0 = 1, y_1 = 4)$$

$$\implies \quad G(z) = \frac{1 - 2z}{1 - 6z + 8z^2} = \frac{1 - 2z}{(1 - 2z)(1 - 4z)}$$

$$= (1 - 4z)^{-1} = \sum_{n=0}^{\infty} 4^n z^n. \quad (2.42)$$

Comparing (2.40) and (2.42), the solution is $y_n = 4^n$.

5. Solve $S(k) - 7S(k - 2) + 6S(k - 3) = 0, S(0) = 8, S(1) = 6$, and $S(2) = 22$.

Solution.

Let the generating function be

$$G(z) = \sum_{k=0}^{\infty} S(k) z^k. \quad (2.43)$$

Multiplying the given equation by z^k and summing for $k \geq 3$, we get

$$\sum_{k=3}^{\infty} S(k) z^k - 7 \sum_{k=3}^{\infty} S(k - 2) z^k + 6 \sum_{k=3}^{\infty} S(k - 3) z^k = 0. \quad (2.44)$$

Consider the first sum

$$\sum_{k=3}^{\infty} S(k) z^k = S(3) z^3 + s(4) z^4 + s(5) z^5 + \ldots$$

$$= G(z) - S(0)S(1)z - S(2)z^2 \quad [\text{using } (2.43)],$$

the second sum

$$\sum_{k=3}^{\infty} s(k-3)z^k = S(0)z^3 + S(1)z^4 + S(2)z^5 + \dots$$

$$= z^3[S(0) + S(1)z + S(2)z^2 + \dots]$$
$$= z^3 G(z) \quad [\text{using } (2.43)],$$

and the third sum

$$\sum_{k=3}^{\infty} s(k-3)z^k = S(0)z^3 + S(1)z^4 + S(2)z^5 + \dots$$

$$= z^3[S(0) + S(1)z + S(2)z^2 + \dots]$$
$$= z^3 G(z) \quad [\text{using } (2.43)].$$

$\therefore$ (2.44) becomes

$$[G(z) - s(0) - s(1)z - s(2)z^2] - 7z^2[G(z) - s(0)] + 6z^3 G(z) = 0.$$

Since $S(0) = 8, S(1) = 6, S(2) = 22$,

$$[G(z) - 8 - 6z - 22z^2] - 7z^2[G(z) - 8] + 6z^3 G(z) = 0$$
$$\implies \quad G(z)[1 - 7z^2 + 6z^3] = 8 + 6z - 34z^2$$
$$\implies \quad G(z) = \frac{8 + 6z - 34z^2}{1 - 7z^2 + 6z^3}$$
$$\implies \quad G(z) = \frac{8 + 6z - 34z^2}{(1-z)(1-2z)(1-3z)}.$$

Using partial fractions,

$$\frac{8 + 6z - 34z^2}{(1-z)(1-2z)(1-3z)} = \frac{A}{1-z} + \frac{B}{1-2z} + \frac{C}{1+3z}$$
$$\implies \quad 8 + 6z - 34z^2 = A(1-2z)(1+3z)$$
$$+ B(1-z)(1+3z) + C(1-z)(1-2z).$$

Put $z = 1 \implies -20 = -4A \implies A = 5$.
Put $z = \dfrac{1}{2} \implies \dfrac{10}{4} = \dfrac{5}{4}B \implies B = 2$.
Put $z = -\dfrac{1}{3} \implies \dfrac{20}{9} = \dfrac{20}{9}C \implies C = 1$.

$$\therefore \quad G(z) = \frac{5}{1-z} + \frac{2}{1-2z} + \frac{1}{1+3z}$$
$$= 5(1-z)^{-1} + 2(1-2z)^{-1} + (1+3z)^{-1}$$
$$= 5\sum_{k=0}^{\infty} z^k + 2\sum_{k=0}^{\infty} (2z)^k + \sum_{k=0}^{\infty} (-3)^k z^k$$

$$= \sum_{k=0}^{\infty} \left[5 + 2^{k+1} + (-3)^k \right] z^k. \tag{2.45}$$

Comparing (2.43) and (2.45), the solution is

$$S(k) = 5 + 2^{k+1} + (-3)^k.$$

6. Suppose that a valid code word is an n-digit number in decimal notation containing an even number of 0's. Let a_n denote the number of valid code words of length n. The sequence $\{a_n\}$ satisfies the recurrence relation $a_n = 8a_{n-1} + 10^{n-1}$ and the initial condition $a_1 = 9$. Use generating function to find an explicit formula for a_n.

Solution.
To make our work with generating function simpler, we extend this sequence by setting $a_0 = 1$ so that $a_1 = 8a_0 + 10^0 = 9$.

Let the generating function be

$$G(z) = \sum_{n=0}^{\infty} a_n z^n. \tag{2.46}$$

Multiplying both sides of $a_n = 8a_{n-1} + 10^{n-1}$ by z^n and summing for all $n \geq 1$, we get

$$\sum_{n=1}^{\infty} a_n z^n = 8 \sum_{n=1}^{\infty} a_{n-1} z^n + \sum_{n=1}^{\infty} 10^{n-1} z^n. \tag{2.47}$$

Consider the first sum

$$\sum_{n=1}^{\infty} a_n z^n = a_1 z + a_2 z^2 + a_3 z^3 + \cdots = G(z) - a_0 \quad \text{[using (2.47)]},$$

the second sum

$$\sum_{n=1}^{\infty} a_{n-1} z^n = a_0 z + a_1 z^2 + a_2 z^3 + \ldots$$

$$= z(a_0 + a_z + a_2 z^2 + \ldots)$$

$$= zG(z) \quad \text{[using (2.46)]},$$

and the third sum

$$\sum_{n=1}^{\infty} 10^{n-1} z^n = z \sum_{n=1}^{\infty} (10z)^n$$

$$= z(1 - 10z)^{-1} = \frac{z}{1 - 10z}.$$

$\therefore$ (2.47) becomes

$$[G(z) - a_0] = 8zG(z) + \frac{z}{1 - 10z}.$$

Using $a_0 = 1$,

$$G(z)[1 - 8z] = 1 + \frac{z}{1 - 10z} = \frac{1 - 9z}{1 - 10z}$$

$$\implies \quad G(z) = \frac{1 - 9z}{(1 - 8z)(1 - 10z)}.$$

Using partial fraction,

$$\frac{1 - 9z}{(1 - 8z)(1 - 10z)} = \frac{A}{1 - 8z} + \frac{B}{1 - 10z}$$

$$\implies \quad 1 - 9z = A(1 - 10z) + B(1 - 8z).$$

$$z = \frac{1}{8} \implies -\frac{1}{8} = A\left(1 - \frac{5}{4}\right)$$

$$\implies -\frac{1}{8} = -\frac{1}{4}A$$

$$\implies A = \frac{1}{2}.$$

$$z = \frac{1}{10} \implies \frac{1}{10} = \frac{1}{5}B$$

$$\implies B = \frac{1}{2}.$$

$$\therefore \quad G(z) = \frac{1}{2}(1 - 8z)^{-1} + \frac{1}{2}(1 - 10z)^{-1}$$

$$= \frac{1}{2}\sum_{n=0}^{\infty}(8z)^n + \frac{1}{2}\sum_{n=0}^{\infty}(10z)^n$$

$$= \sum_{n=0}^{\infty}\frac{1}{2}(8^n + 10^n)\,z^n. \tag{2.48}$$

Comparing (2.46) and (2.48), the solution is

$$a_n = \frac{1}{2}(8^n + 10^n).$$

7. Solve $S(n) - 2S(n-1) - 3S(n-2) = 0$, $n \geq 2$ with $S(0) = 3$ and $S(1) = 1$ using generating function.

Solution.

Let the generating function be

$$G(z) = \sum_{n=0}^{\infty} S(n)z^n. \tag{2.49}$$

Multiplying the given equation by z^n and summing for $n \geq 2$, we get

$$\sum_{n=2}^{\infty} S(n)z^n - 2\sum_{n=2}^{\infty} S(n-1)z^n - 3\sum_{n=2}^{\infty} S(n-2)z^n = 0. \quad (2.50)$$

Consider the first sum

$$\sum_{n=2}^{\infty} S(n)z^n = S(2)z^2 + S(3)z^3 + S(4)z^4 + \ldots$$

$$= G(z) - S(0) - S(1)z \quad [\text{using } (2.49)],$$

the second sum

$$\sum_{n=2}^{\infty} S(n-1)z^n = S(1)z^2 + S(2)z^3 + S(3)z^4 + \ldots$$

$$= z[S(1)z + S(2)z^2 + S(3)z^3 + \ldots]$$
$$= z[G(z) - S(0)] \quad [\text{using } (2.49)],$$

and the third sum

$$\sum_{n=2}^{\infty} S(n-2)z^n S(0)z^2 + S(1)z^3 + S(2)z^4 + \ldots$$

$$= z^2[S(0) + S(1)z + S(2)z^2 + \ldots]$$
$$= z^2 G(z).$$

$\therefore \quad$ (2.50) becomes

$$[G(z) - S(0) - S(1)z] - 2z[G(z) - S(0)] - 3z^2 G(z) = 0.$$

Using $S(0) = 3$ and $S(1) = 1$, we get

$$[G(z) - 3 - z] - 2z[G(z) - 3] - 3z^2 G(z) = 0$$
$$\implies \quad G(z)[1 - 2z - 3z^2] = 3 - 5z$$
$$\implies \quad G(z) = \frac{3 - 5z}{1 - 2z - 3z^2} = \frac{3 - 5z}{(1 - 3z)(1 + z)}.$$

Using partial fractions,

$$\frac{3 - 5z}{(1 - 3z)(1 + z)} = \frac{A}{1 - 3z} + \frac{B}{1 + z}$$
$$\implies \quad 3 - 5z = A(1 + z) + B(1 - 3z).$$

Put $z = -1 \implies 8 = 4B \implies B = 2.$

Put $z = \dfrac{1}{3} \Longrightarrow \dfrac{4}{3} = \dfrac{4}{3}A \Longrightarrow A = 1.$

$$\therefore \quad G(z) = \frac{1}{1 - 3z} + \frac{2}{1 + z}$$
$$= (1 - 3z)^{-1} + 2(1 + z)^{-1}$$
$$= \sum_{n=0}^{\infty} (3z)^n + 2 \sum_{n=0}^{\infty} (-z)^n$$
$$= \sum_{n=0}^{\infty} \left[3^n + 2(-1)^n \right] z^n. \qquad (2.51)$$

Comparing (2.49) and (2.51), the solution is

$$S(n) = 3^n + 2(-1)^n.$$

2.8.4 Problems for Practice

1. Find the generating function of the following sequences.

   (i) $2, 2, 2, 2, 2, 2, \ldots$

   (ii) $a_n = (-2)^n$

   (iii) $0, 0, 1, 1, 1, 1, \ldots$

   (iv) $1, 0, -1, 0, 1, 0, 0, -1, 0, 1, 0, -1, 0, \ldots$

2. For the following expressions, identify the sequences having the expression as a generating function. [Hint: Use partial fractions.]

   (i) $\dfrac{5 + 2z}{1 - 4z^2}$

   (ii) $\dfrac{6 - 29z}{1 - 11z + 30z^2}$

   (iii) $\dfrac{32 - 22z}{2 - 3z + z^2}$

   (iv) $\dfrac{3 + 7z}{1 + 3z - 4z^2}$

   (v) $\dfrac{3 + 5z}{1 - 2z - 3z^2}.$

3. Find the generating functions for the following sequences satisfying the given initial conditions.

   (i) $y_{n+2} + 2y_{n+1} - 15y_n = 0$; $y_0 =, y_1 = 1.$

   (ii) $y_{n+1} - y_{n-1} = 0$; $y_0 = 0, y_1 = 1.$

   (iii) $y_{n+2} - 2y_{n+1} - 3y_n = 0$; $y_0 = 0, y_1 = 1.$

   (iv) $S(n) - 2S(n-1) - 3S(n-2) = 0$; $S(0) = 3, S(1) = 1.$

   (v) $S(k) + 3S(k-1) - S(k-2) = 0$; $S(0) = 3, S(1) = 2.$

4. Using generating function, solve the following recurrence relations.

(i) $y_n = 7y_{n-1}; y_0 = 5$.

(ii) $y_n = 3y_{n-1} + 4^{n-1}; y_0 = 1$.

(iii) $y_n + 2y_{n-1} - 15y_{n-2} = 0; y_0 = 0, y_1 = 1$.

(iv) $y_{n+2} - 8y_{n+1} + 16y_n = 0; y_0 = 0, y_1 = 8$.

(v) $y_{n+2} - 2y_{n+1} + y_n = 0; y_0 = 2, y_1 = 1$.

(vi) $y_{n+2} - y_{n+1} - 6y_n = 0; y_0 = 2, y_1 = 1$.

(vii) $y_{n+2} - 5y_{n+1} + 6y_n = 0; y_0 = 1, y_1 = 3$.

5. Use a generating function to find an explicit formula for the Fibonacci numbers.

2.9 Inclusion–Exclusion Principle

Let A and B be any finite sets. Then,

$$n(A \cup B) = n(A) + n(B) - n(A \cap B).$$

In other words, to find the number $n(A \cup B)$ of elements in the union $A \cup B$, we add $n(A)$ and $n(B)$, and then we subtract $n(A \cap B)$; that is, we "include" $n(A)$ and $n(B)$, and we "exclude" $n(A \cap B)$. This follows from the fact that, when we add $n(A)$ and $n(B)$, we have counted the elements of $A \cap B$ twice. This principle holds for any number of sets. We first state it for three sets.

Theorem 2.9.1 *For any finite sets A, B, and C, we have*

$$n(A \cup B \cup C)$$
$$= n(A) + n(B) + n(C) - n(A \cap B)$$
$$- n(B \cap C) - n(A \cap C) + n(A \cap B \cap C).$$

That is, we "include" $n(A)$, $n(B)$, $n(C)$, we "exclude" $n(A \cap B)$, $n(B \cap C)$, $n(A \cap C)$, and we "include" $n(A \cap B \cap C)$.

Example:
Find the number of mathematics students at a college taking at least one of the languages French, German, and Russian given the following data:

65 study French, 20 study French and German,

45 study German, 25 study French and Russian,

42 study Russian, 15 study German and Russian,

and 8 study all three languages.

We want to find $n(F \cup G \cup R)$, where F, G, and R denote the sets of students studying French, German, and Russian, respectively.

By the inclusion–exclusion principle,

$$n(F \cup G \cup R) = n(F) + n(G) + n(R) - n(F \cap G)$$
$$- n(F \cap R) - n(G \cap R) + n(F \cap G \cap R)$$
$$= 65 + 45 + 42 - 20 - 25 - 15 + 8$$
$$= 100.$$

∴   100 students study at least one of the languages.

Note:

Principle of inclusion–exclusion can also be denoted as

(i) $|A \cup B| = |A| + |B| - |A \cap B|$ if A and B are not disjoint sets

(ii) $|A \cup B| = |A| + |B|$ if A and B are disjoint sets

where $|A| = n(A) =$ cardinality of $A =$ number of elements in A.

2.9.1   Solved Problems

1. Find the number of positive integers not exceeding 100 that are divisible by 7 or by 11.

   Solution.

   Let A be the set of positive integers not exceeding 100 that are divisible by 7.

   Let B be the set of positive integers not exceeding 100 that are divisible by 11.

   Then, $A \cup B$ is the set of positive integers not exceeding 100 that are divisible by either 7 or 11, and $A \cap B$ is the set of positive integers not exceeding 100 that are divisible by both 7 and 11.

   We know that among the positive integers not exceeding 100, there are $\left\lfloor \dfrac{100}{7} \right\rfloor$ integers divisible by 7 and $\left\lfloor \dfrac{100}{11} \right\rfloor$ integers divisible by 11.

   Since 7 and 11 are relatively prime, the integers divisible by both 7 and 11 are those divisible by 7 and 11.

   There are $\left\lfloor \dfrac{100}{7 \times 11} \right\rfloor$ positive integers not exceeding 100 that are divisible by both 7 and 11.

$$\therefore \quad |A \cap B| = |A| + |B| - |A \cap B|$$
$$= \left[\frac{100}{7}\right] + \left[\frac{100}{11}\right] - \left[\frac{100}{7 \times 11}\right]$$
$$= 14 + 9 - 1 = 22.$$

2. Among the first 1000 positive integers, determine the integers which are not divisible by 5, nor by 7, nor by 9.

Solution.

Let A = set of integers divisible by 5

B = set of integers divisible by 7

C = set of integers divisible by 9.

$$\therefore \quad |A| = \left[\frac{1000}{5}\right] = 200; \quad |B| = \left[\frac{1000}{7}\right] = 142; \quad C = \left[\frac{1000}{9}\right] = 111.$$

$$|A \cap B| = \left[\frac{1000}{LCM(5,7)}\right] = \left[\frac{1000}{35}\right] = 28$$

$$|B \cap C| = \left[\frac{1000}{LCM(7,9)}\right] = \left[\frac{1000}{63}\right] = 15$$

$$|A \cap C| = \left[\frac{1000}{LCM(5,9)}\right] = \left[\frac{1000}{45}\right] = 22$$

$$|A \cap B \cap C| = \left[\frac{1000}{LCM(5,7,9)}\right] = \left[\frac{1000}{5 \times 7 \times 9}\right] = \left[\frac{1000}{315}\right] = 3.$$

The number of integers divisible by 5, 7, and 9 is

$$|A \cup B \cup C| = |A| + |B| + |C| - |A \cap B| - |B \cap C| - |A \cap C| + |A \cap B \cap C|$$
$$= 200 + 142 + 111 - 28 - 15 - 22 + 3$$
$$= 391.$$

The number of integers not divisible by 5 nor 7 nor 9

$$= \text{Total number of integers - integers divisible by 5, 7, and 9}$$
$$= 1000 - 391$$
$$= 609.$$

3. In a survey of 300 students, 64 had taken a Mathematics course, 94 had taken an English course, 58 had taken a Computer course, 28 had taken both Mathematics and Computer courses, 26 had taken both English and Mathematics courses, 22 had taken both

English and Computer courses, and 14 had taken all three courses. How many students were surveyed who had taken none of the three courses?

Solution.

Given: $|M| = 64;$ $|E| = 94;$ $|C| = 58;$

$|M \cap C| = 28;$ $|M \cap E| = 26;$ $|E \cap C| = 22;$ $|M \cap C \cap E| = 14.$

$$
\begin{aligned}
|M \cup E \cup E| &= |M| + |E| + |C| - |M \cap E| \\
&\quad - |M \cap C| - |E \cap C| + |M \cap E \cap C| \\
&= 64 + 94 + 58 - 26 - 28 - 22 + 14 \\
&= 154.
\end{aligned}
$$

$\therefore$ Students who had taken none of the courses $= 300 - 154 = 146.$

4. How many solutions does $x_1 + x_2 + x_3 = 13$ have, where x_1, x_2, and x_3 are non-negative integers with $x_1 < 6$, $x_2 < 6$, and $x_3 < 6$?

Solution.

To apply the principle of inclusion–exclusion, let a solution have property P_1 if $x \geq 6$, property P_2 if $x_2 \geq 6$, and property P_3 if $x_3 \geq 6$. The number of solutions satisfying the inequalities $x_1 < 6$, $x_2 < 6$, and $x_3 < 6$ is

$$
\begin{aligned}
N(P_1' P_2' P_3') &= N - N(P_1) - N(P_2) - N(P_3) + N(P_1 P_2) \\
&\quad + N(P_1 P_3) + N(P_2 P_3) - N(P_1 P_2 P_3) \qquad (2.52)
\end{aligned}
$$

where $N = $ total number of solutions

$$
= C(3 + 13 - 1, 13) = C(15, 13) = 105.
$$

$N(P_1) = $ number of solutions with $x_1 \geq 6$

$\qquad = C(3 + 7 - 1, 7) = C(9, 7) = 36$

$N(P_2) = $ number of solutions with $x_2 \geq 6$

$\qquad = C(3 + 7 - 1, 7) = C(9, 7) = 36$

$N(P_3) = $ number of solutions with $x_3 \geq 6$

$\qquad = C(3 + 7 - 1, 7) = C(9, 7) = 36$

$N(P_1 P_2) = $ number of solutions with $x_1 \geq 6$ and $x_2 \geq 6$

$\qquad = C(3 + 1 - 1, 1) = C(3, 1) = 3$

$N(P_1 P_3) = $ number of solutions with $x_1 \geq 6$ and $x_3 \geq 6$

$\qquad = C(3 + 1 - 1, 1) = C(3, 1) = 3$

$N(P_2 P_3) = $ number of solutions with $x_2 \geq 6$ and $x_3 \geq 6$

$\qquad = C(3 + 1 - 1, 1) = C(3, 1) = 3$

$N(P_1 P_2 P_3) = $ number of solutions with $x \geq 6, x_2 \geq 6,$ and $x_3 \geq 6 = 0.$

Inserting these quantities into the formula $N(P_1'P_2'N_3')$ shows that the number of solutions with $x_1 \leq 6$, $x_2 \leq 6$, and $x_3 \leq 6$ equals [implying from (2.52)]

$$N(P_1'P_2'N_3') = 105 - 36 - 36 - 36 + 3 + 3 + 3 - 0 = 6.$$

5. A survey of 500 students from a school produced the following information. 200 play volleyball, 120 play hockey, 60 play both volleyball and hockey. How many are not playing either volleyball or hockey?

Solution.

Let A be the set of students who play volleyball.

Let B be the set of students who play hockey.

Given: $|A| = 200$, $|B| = 120$, $|A \cap B| = 60$.

By the principle of inclusion–exclusion, the number of students playing either volleyball or hockey is

$$|A \cup B| = |A| + |B| - |A \cap B|$$
$$= 200 + 120 - 60$$
$$= 260.$$

∴ The number of students not playing either volleyball or hockey

$$= 500 - 260 = 240.$$

6. A total of 1232 students have taken a course in Russian, 879 have taken a course in German, and 114 have taken a course in French. Further 103 have taken a course in both Russian and German, 23 have taken a course in Russian and French, and 14 have taken a course in German and French. If 2092 students have taken at least one of the courses Russian, German, and French, how many students have taken a course in all three languages.

Solution.

Let A be the set of students who have taken a course in Russian.

Let B be the set of students who have taken a course in German.

Let C be the set of students who have taken a course in French.

Given: $|A| = 1232$, $|B| = 879$, $|C| = 114$,

$|A \cap B| = 103$, $|A \cap C| = 23$, $|B \cap C|14$, $|A \cap B \cap C| = 2092$.

By the principle of inclusion–exclusion, we have

$$|A \cup B \cup C| = |A| + |B| + |C| - |A \cap B| - |B \cap C|$$
$$- |A \cap C| + |A \cap B \cap C|$$
$$\implies \quad 2092 = 1232 + 879 + 114 - 103 - 23 - 14 + |A \cap B \cap C|$$
$$\implies \quad |A \cap B \cap C| = 7.$$

∴ There are seven students who have taken a course in Russian, German, and French.

7. In a survey of 100 students, it was found that 30 studied Mathematics, 54 studied Statistics, 25 studied Operations Research, 1 studied all the three subjects, 20 studied Mathematics and Statistics, 3 studied Mathematics and Operations Research, and 15 studied Statistics and Operations Research.

(i) How many students studied none of these subjects?

(ii) How many students studied only Mathematics?

Solution.

Let A denote the set of students who studied Mathematics.

Let B denote the set of students who studied Statistics.

Let C denote the set of students who studied Operations Research.

Given: $|A| = 30$, $|B| = 54$, $C = 25$,

$|A \cap B| = 20$, $|A \cap C| = 3$, $|B \cap C| = 15$, $|A \cap B \cap C| = 1$.

(i) By the principle of inclusion–exclusion, the number of students who studied any one of the subjects is

$$|A \cup B \cup C| = |A| + |B| + |C| - |A \cap B| - |B \cap C|$$
$$- |A \cap C| + |A \cap B \cap C|$$
$$= 30 + 54 + 25 - 2003 - 15 + 1 = 72.$$

∴ Number of students who studied none of these subjects $= 100 - 72 = 28$.

(ii) Number of students who studied only Mathematics and Statistics

$$= |A \cap B| - |A \cap B \cap C| = 20 - 1 = 19.$$

Number of students who studied only Mathematics and Operations Research

$$= |A \cap C| - |A \cap B \cap C| = 3 - 1 = 2.$$

∴ Number of students who studied only Mathematics $= 30 - 19 - 2 - 1 = 8$.

8. How many positive integers not exceeding 1000 are divisible by 7 or 11?

Solution.

Let A denote the set of positive integers not exceeding 1000 that are divisible by 7.

Let B denote the set of positive integers not exceeding 1000 that are divisible by 11. Then,

$$|A| = \left[\frac{1000}{7}\right] = 142, |B| = \left[\frac{1000}{11}\right] = 90, |A \cap B| = \left[\frac{1000}{7 \times 11}\right] = 12.$$

The number of positive integers not exceeding 1000 that are divisible by either 7 or 11 is $|A \cup B|$.

By the principle of inclusion–exclusion,

$$|A \cup B| = |A| + |B| - |A \cap B|$$
$$= 142 + 90 - 12$$
$$= 220.$$

9. Determine n such that $1 \le n \le 100$ and it is not divisible by 5 or 7.

   Solution.

   Let A denote the number n, $1 \le n \le 100$, which is divisible by 5.

   Let B denote the number n, $1 \le n \le 100$, which is divisible by 7.

   Then,

   $$|A| = \left[\frac{100}{5}\right] = 20, |B| = \left[\frac{100}{7}\right] = 14, |A \cap B| = \left[\frac{100}{5 \times 5}\right] = 2.$$

   By the principle of inclusion–exclusion, the number n, $1 \le n \le 100$, which is divisible by either 5 or 7 is $|A \cup B|$.

   $$|A \cup B| = |A| + |B| - |A \cap B|$$
   $$= 20 + 14 - 2$$
   $$= 32.$$

   $\therefore$ The number n, $1 \le n \le 100$, which is not divisible by 5 and 7 is

   $$= 100 - 32 = 68.$$

10. A survey among 100 students shows that of the three ice cream flavours vanilla, chocolate, and strawberry, 50 students like vanilla, 43 like chocolate, 28 like strawberry, 13 like vanilla and chocolate, 11 like chocolate and strawberry, 12 like strawberry and vanilla, and 5 like all of them. Find the number of students surveyed who like the following flavours:

    (i) chocolate but not strawberry
    (ii) chocolate and strawberry but not vanilla
    (iii) vanilla or chocolate but not strawberry.

    Solution.

    Let A denote the set of students who like vanilla.

    Let B denote the set of students who like chocolate.

    Let C denote the set of students who like strawberry.

    Since five students like all flavours, $|A \cap B \cap C| = 5$.

Twelve students like both strawberry and vanilla $\implies$ $|A \cap C| = 12$.

But five of them like chocolate also $\implies$ $|A \cap C| - 5 = 12 - 5 = 7$.

Six of them like vanilla $\implies$ $|B \cap C| - 6 = 12 - 6 = 6$.

Out of 28 students who like strawberry,

we have already accounted for $7 + 5 + 6 = 18$.

$\therefore$ The remaining ten students belong to the set $C - (A \cup B)$.

Similarly,

$$|A - (B \cup C)| = 30 \qquad \text{and} \qquad |B - (A \cup C)| = 24.$$

Hence, we have accounted for 90 of the 100 students. The remaining ten students like outside the region $A \cup B \cup C$. Now,

(i) $|B - C| = 24 + 8 = 32$.

$\therefore$ 32 students like chocolate but not strawberry.

(ii) $|(B \cap C) - A| = 6$.

$\therefore$ Six students like both chocolate and strawberry but not vanilla.

(iii) $|(A \cup B) - C| = 30 + 8 + 24 = 62$.

$\therefore$ 62 students like vanilla or chocolate, but not strawberry.

11.   Find the number of integers between 1 and 250 that are not divisible by any of the integers 2, 3, 5, and 7.

Solution.

Let A denote the set of integers between 1 and 250 that are divisible by 2.

Let B denote the set of integers between 1 and 250 that are divisible by 3.

Let C denote the set of integers between 1 and 250 that are divisible by 5.

Let D denote the set of integers between 1 and 250 that are divisible by 7.

Now,

$$|A| = \left[\frac{250}{2}\right] = 125, \qquad |B| = \left[\frac{250}{3}\right] = 83$$

$$|C| = \left[\frac{250}{5}\right] = 5, \qquad |D| = \left[\frac{250}{7}\right] = 35.$$

Number of integers between 1 and 250 that are divisible by 2 and 3

$$= |A \cap B| = \left[\frac{250}{2 \times 3}\right] = 41$$

Number of integers between 1 and 250 that are divisible by 2 and 5

$$= |A \cap C| = \left[\frac{250}{2 \times 5}\right] = 25.$$

Similarly,

$$|A \cap D| = \left[\frac{250}{2 \times 7}\right] = 17$$

$$|B \cap C| = \left[\frac{250}{3 \times 5}\right] = 16$$

$$|B \cap D| = \left[\frac{250}{3 \times 7}\right] = 11$$

$$|C \cap D| = \left[\frac{250}{5 \times 7}\right] = 7.$$

Number of integers between 1 and 250 that are divisible by 2, 3, and 5

$$= |A \cap B \cap C| = \left[\frac{250}{2 \times 3 \times 50}\right] = 8.$$

Similarly,

$$|A \cap B \cap D| = \left[\frac{250}{2 \times 3 \times 7}\right] = 5$$

$$|A \cap C \cap D| = \left[\frac{250}{2 \times 5 \times 7}\right] = 3$$

$$|B \cap C \cap D| = \left[\frac{250}{3 \times 5 \times 7}\right] = 2$$

$$|A \cap B \cap C \cap D| = \left[\frac{250}{2 \times 3 \times 5 \times 7}\right] = 1.$$

The number of integers between 1 and 250 that are divisible by 2, 3, 5, and 7 is $|A \cup B \cup C \cup D|$.

By principle of inclusion–exclusion,

$$
\begin{aligned}
|A \cup B \cup C \cup D| = {}& |A| + |B| + |C| + |D| - |A \cap B| - |B \cap C| \\
& - |C \cap D| - |A \cap C| - |B \cap D| - |A \cap D| \\
& + |A \cap B \cap C| + |A \cap B \cap D| + |A \cap C \cap D| \\
& + |B \cap C \cap D| - |A \cap B \cap C \cap D| \\
= {}& 125 + 83 + 50 + 35 - 41 - 25 - 17 - 16 - 11 - 7 \\
& + 8 + 5 + 3 + 2 - 1 \\
= {}& 193.
\end{aligned}
$$

$\therefore$ Number of integers between $\left.\begin{array}{l}\text{1 and 250 that are not}\\\text{divisible by 2, 3, 5, and 7}\end{array}\right\} = 250 - 193 = 57.$

12. A survey shows that 57% of Indians like coffee whereas 75% like tea. What can you say about the percentage of Indians who like both coffee and tea.

    Solution.

    Let A denote the set of Indians who like coffee.

    Let B denote the set of Indians who like tea.

    Assume the total population is 100.

    $\therefore$ $|A| = 57;$ $|B| = 75.$ Now,

    $$\begin{aligned}|A \cup B| &= |A| + |B| - |A \cap B|\\ &= 57 + 75 - |A \cap B|\\ &= 132 - |A \cap B|.\end{aligned}$$

    Since $|A \cup B| \leq 100$, it follows that

    $$|A \cap B| \geq 32. \tag{2.53}$$

    Since $A \cap B \subseteq A$ and $A \cap B \subseteq B$, we have

    $$|A \cap B| \leq |A| \quad \text{and} \quad |A \cap B| \leq |B|.$$

    $\therefore$ $|A \cap B| \leq 57$ and $|A \cap B| \leq 75.$

    $\Longrightarrow$ $|A \cap B| \leq 57. \tag{2.54}$

    From (2.53) and (2.54), the percentage of Indians who like both coffee and tea lies between 32 and 57.

13. Out of 100 students in a college, 38 play tennis, 57 play cricket, 31 play hockey, 9 play cricket and hockey, 10 play hockey and tennis, and 12 play tennis and cricket. How many play

    (i) all three games

    (ii) just one game

    (iii) tennis and cricket but not hockey.

    Assume that each student plays at least one game.

    Solution.

    Let $T, C,$ and H denote the set of students playing tennis, cricket, and hockey, respectively (Figure 2.1).

    Given: $|T| = 38,$ $|C| = 57,$ $|H| = 31$

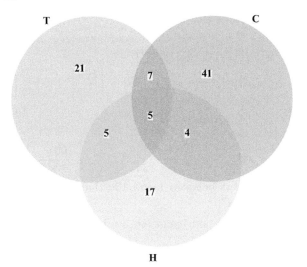

FIGURE 2.1
Venn diagram

$|T \cap C| = 12, \quad |T \cap H| = 10, \quad |C \cap H| = 9, \quad |T \cup C \cup H| = 100.$

Number of students who play all three games $= |T \cap C \cap H|$.

By principle of inclusion–exclusion, we have

$$|T \cap C \cap H| = |T| + |C| + |H| - |T \cap C| - |C \cap H|$$
$$- |T \cap H| + |T \cap C \cap H|$$
$$\implies \quad 100 = 38 + 57 + 31 - 12 - 9 - 10 + |T \cap C \cap H|$$
$$\implies \quad |T \cap C \cap H| = 100 - 126 + 31 = 5.$$

$\therefore$ Number of students who play all three games $= 5$.

From the given data, we have

Number of students

playing just one game = Number of students playing tennis only

+ Number of students playing cricket only

+ Number of students playing hockey only

$$= 21 + 41 + 17 = 79.$$

Number of students playing tennis and cricket but not hockey

$$= |T \cap C| - |T \cap C \cap H| = 12 - 5 = 7.$$

14. How many integers between 1 and 100 are

   (i) not divisible by 7, 11, or 13

   (ii) divisible by 3 but not by 7?

Solution.

Let A, B, and C be the set of integers between 1 and 100 that are divisible by 7, 11, and 13, respectively.

$$\therefore \qquad |A| = \left[\frac{100}{7}\right] = 14; \quad |B| = \left[\frac{100}{11}\right]; \quad |C| = \left[\frac{100}{13}\right] = 7;$$

$$|A \cap B| = \left[\frac{100}{7 \times 11}\right] = 1; \quad |A \cap C| = \left[\frac{100}{7 \times 13}\right] = 1;$$

$$|B \cap C| = \left[\frac{100}{11 \times 13}\right] = 0; \quad \left[\frac{100}{7 \times 11 \times 13}\right] = 0.$$

Number of integers between 1 and 100 that are divisible by 7, 11, or 13 is $|A \cup B \cup C|$.

By principle of inclusion–exclusion, we have

$$|A \cup B \cup C| = |A| + |B| + |C| - |A \cap B| - |B \cap C|$$
$$- |A \cap C| + |A \cap B \cap C|$$
$$= 14 + 9 + 7 - 1 - 0 - 1 + 0$$
$$= 28.$$

(i) Number of integers between 1 and 100 that are not divisible by 7, 11, or 13 $= 100 - 28 = 72$.

(ii) Let U and V denote the set of integers between 1 and 100 that are divisible by 3 and 7, respectively.

$$|U| = \left[\frac{100}{3}\right] = 33; \quad |V| = \left[\frac{100}{7}\right] = 14; \quad |U \cap V| = \left[\frac{100}{3 \times 7}\right] = 4.$$

$$\therefore \quad \text{Number of integers divisible by 3 but not by } 7 = |U| - |U \cap V|$$
$$= 33 - 4$$
$$= 29.$$

15. Find the number of integers between 1 and 100 that are divisible by

(i) 2, 3, 5, or 7

(ii) 2, 3, 5 but not by 7.

Solution.

Let A, B, C, and D denote the set of positive integers between 1 and 100 that are divisible by 2, 3, 5, and 7, respectively.

$$\therefore \quad |A| = \left[\frac{100}{2}\right] = 50; \qquad |B| = \left[\frac{100}{3}\right] = 33;$$

$$|C| = \left[\frac{100}{5}\right] = 20; \qquad |D| = \left[\frac{100}{7}\right] = 14;$$

$$|A \cap B| = \left[\frac{100}{2 \times 3}\right] = 16; \qquad |A \cap C| = \left[\frac{100}{2 \times 5}\right] = 10;$$

$$|A \cap D| = \left[\frac{100}{2 \times 7}\right] = 7; \qquad |B \cap C| = \left[\frac{100}{3 \times 5}\right] = 7;$$

$$|B \cap D| = \left[\frac{100}{3 \times 7}\right] = 4; \qquad |C \cap D| = \left[\frac{100}{5 \times 7}\right] = 2;$$

$$|A \cap B \cap C| = \left[\frac{100}{2 \times 3 \times 5}\right] = 3; \quad |A \cap B \cap D| = \left[\frac{100}{2 \times 3 \times 7}\right] = 2;$$

$$|A \cap C \cap D| = \left[\frac{100}{2 \times 5 \times 7}\right] = 1; \quad |B \cap C \cap D| = \left[\frac{100}{3 \times 5 \times 7}\right] = 0;$$

$$|A \cap B \cap C \cap D| = \left[\frac{100}{2 \times 3 \times 5 \times 7}\right] = 0.$$

(i) By the principle of inclusion–exclusion, we have

$$
\begin{aligned}
|A \cup & B \cup C \cup D| \\
&= |A| + |B| + |C| + |D| - |A \cap B| - |A \cap C| - |A \cap D| \\
&\quad - |B \cap C| - |B \cap D| + |A \cap B \cap C| + |A \cap B \cap D| \\
&\quad + |A \cap C \cap D| + |B \cap C \cap D| - |A \cap B \cap C \cap D| \\
&= 50 + 33 + 20 + 14 - 16 - 10 - 7 - 7 - 4 - 2 \\
&\quad + 3 + 2 + 1 + 0 - 0 \\
&= 117 - 46 + 6 \\
&= 123 - 46 \\
&= 77.
\end{aligned}
$$

(ii) The number of integers between 1 and 100 that are divisible by 2, 3, 5 but not by 7
$$= |A \cap B \cap C| - |A \cap B \cap C \cap D| = 3 - 0 = 3.$$

16. How many prime numbers not exceeding 100 are there? Or determine a prime number n, where $1 \le n \le 100$.

Solution.

To find the number of primes not exceeding 100, first note that a composite integer not exceeding 100 must have a prime factor not exceeding 10.

The primes not exceeding 100 are 2, 3, 5, and 7 and the numbers that are divisible by none of 2, 3, 5, or 7.

Let P_1 be the property that an integer is divisible by 2.

Let P_2 be the property that an integer is divisible by 3.

Let P_3 be the property that an integer is divisible by 5.

Let P_4 be the property that an integer is divisible by 7.

Number of primes not exceeding $= 4 + N(P_1' P_2' P_3' P_4')$.

Now,

$$N(P_1'P_2'P_3'P_4') = 99 - N(P_1) - N(P_2) - N(P_3) - N(P_4)$$
$$+ N(P_1P_2) + N(P_1P_3) + N(P_1P_4)$$
$$+ N(P_2P_3) + N(P_2P_4) + N(P_3P_4)$$
$$- N(P_1P_2P_3) - N(P_1P_2P_4) - N(P_1P_3P_4)$$
$$- N(P_2P_3P_4) - N(P_1P_2P_3P_4)$$

($\because$ there are 99 integers > 1 and not exceeding 100).

$$N(P_1'P_2'P_3'P_4') = 99 - \left[\frac{100}{2}\right] - \left[\frac{100}{3}\right] - \left[\frac{100}{5}\right] - \left[\frac{100}{7}\right]$$
$$+ \left[\frac{100}{2 \times 3}\right] + \left[\frac{100}{2 \times 5}\right] + \left[\frac{100}{2 \times 7}\right]$$
$$+ \left[\frac{100}{3 \times 5}\right] + \left[\frac{100}{3 \times 7}\right] + \left[\frac{100}{5 \times 7}\right]$$
$$- \left[\frac{100}{2 \times 3 \times 5}\right] - \left[\frac{100}{2 \times 3 \times 7}\right] - \left[\frac{100}{2 \times 5 \times 7}\right]$$
$$- \left[\frac{100}{3 \times 5 \times 7}\right] + \left[\frac{100}{2 \times 3 \times 5 \times 7}\right]$$
$$= 99 - 50 - 33 - 20 - 14 + 16 + 10 + 7 + 6$$
$$+ 4 + 2 - 3 - 2 - 1 - 0 + 0$$
$$= 21.$$

$\therefore$ There are $4 + 21 = 25$ primes.

17. How many solutions does $x_1 + x_2 + x_3 = 11$ have, where x_1, x_2, and x_3 are non-negative integers with $x_1 \leq 3$, $x_2 \leq 4$, and $x_3 \leq 6$?

Solution.

Let P_1 be the property that $x_1 > 3$.

Let P_2 be the property that $x_2 > 4$.

Let P_3 be the property that $x_3 > 6$.

Now, the number of solutions satisfying the inequalities $x_1 \leq 1$, $x_2 \leq 4$, and $x_3 \leq 6$ is $N(P_1'P_2'P_3')$.

By principle of inclusion–exclusion, we have

$$N(P_1'P_2'P_3')N - N(P_1) - N(P_2) - N(P_3)$$
$$+ N(P_1P_2) + N(P_1P_3) + N(P_2P_3) - N(P_1P_2P_3).$$

Now,

$$N = \text{Total number of solutions}$$
$$= C(3 + 11 - 1, 11) = 78$$

[since the number of r-combinations from a set with n elements when repetitions are allowed is $(n+r-1)C_r$ ways or $C(n+r-1,r)$].

$N(P_1)$ = Number of solutions with $x_1 \geq 4$
$= C(3+7-1,7) = C(9,7) = 36.$
$N(P_2)$ = Number of solutions with $x_2 \geq 5$
$= C(3+6-1,6) = C(8,6) = 28.$
$N(P_3)$ = Number of solutions with $x_3 \geq 7$
$= C(3+4-1,4) = C(6,4) = 15.$
$N(P_1 P_2)$ = Number of solutions with $x_1 \geq 4$ and $x_2 \geq 5$
$= C(3+2-1,2) = C(4,2) = 6.$
$N(P_1 P_3)$ = Number of solutions with $x_1 \geq 4$ and $x_3 \geq 7$
$= C(3+0-1,0) = C(2,0) = 1.$
$N(P_2 P_3)$ = Number of solutions with $x_2 \geq 5$ and $x_3 \geq 7 = 0.$
$N(P_1 P_2 P_3)$ = Number of solutions with $x_1 \geq 4$, $x_2 \geq 5$ and $x_3 \geq 7 = 0.$

$\therefore \quad N(P_1' P_2' P_3') = 78 - 36 - 28 - 15 + 6 + 1 + 0 + 0 = 6.$

Hence, the equation $x_1 + x_2 + x_3 = 11$ with respect to the given conditions has six solutions.

2.9.2 Problems for Practice

1. Find the number of elements in $A_1 \cup A_2 \cup A_3$ if there are 100 elements in each set if

   (i) the sets are pairwise disjoint
   (ii) there are 50 common elements in each pair of sets and no element in all three sets
   (iii) the sets are equal.

2. How many elements are in the union of four sets if the sets have 50, 60, 70, and 80 elements, respectively, each pair of sets has five elements in common, each trio of the sets has one common element, and no element is in all four sets?

3. How many bit strings of length 8 do not contain six consecutive 0's?

4. How many solutions does the equation $x_1 + x_2 + x_3 = 16$ have where $x_1, x_2,$ and x_3 are non-negative integers less than 6?

5. How many ways are there to distribute six different toys to three different children such that each child gets at least one toy?

6. How many ways can the digits 0, 1, 2, 3, 4, 5, 6, 7, 8, 9 be arranged so that no even digit is in its original position?

7. By using principle of inclusion–exclusion, find how many positive integers between 1 and 250 are

   (i) divisible by 2, 3, 5, or 7

   (ii) not divisible by 2, 3, 5, and 7.

8. How many positive integers not exceeding 1000 are divisible by 7 or 11?

9. How many elements are in $A_1 \cup A_2$ if there are 12 elements in A_1, 18 elements in A_2, and

   (i) $|A_1 \cap A_2| = 1$

   (ii) $|A_1 \cap A_2| = 6$?

10. There are 2504 Computer Science students at a college. Of these, 1876 have taken a course in Pascal, 999 have taken a course in Fortran, and 345 have taken course in C. Further, 876 have taken courses in both Pascal and Fortran, 231 have taken courses in both Fortran and C, and 290 have taken courses in both Pascal and C. If 189 of these students have taken courses in Fortran, Pascal, and C, how many of these 2504 students have not taken a course in any of these 3 programming languages?

11. How many permutations of ten digits either begin with the three digits 987, contain the digits 45 in the fifth and sixth positions, or end with the three digits 123?

12. Find the probability that when four numbers from 1 to 100, inclusive, are picked at random with no repetitions allowed, either all are odd, all are divisible by 3, or all are divisible by 5.

13. Find the number of primes less than 200 using the principle of inclusion–exclusion.

14. How many derangements are there of a set with seven elements?

15. How many positive integers less than 10000 are not the second or higher powers of an integer?

16. How many integers between 1 and 300 are divisible by

   (i) at least one of 3, 5, 7

   (ii) 3 and 5 but not by 7

   (iii) 5 but not by 3 and 7.

17. A survey of households in the United States reveals that 96% have at least one television set, 98% have telephone service, and 95% have telephone service and at least one television set. What percentage of households in the United States have neither telephone service nor a television set?

18. How many students are enrolled in a course either in calculus, discrete mathematics, data structures, or programming languages at a school if there are 507, 292, 312, and 344 students in these courses, respectively; 14 in both calculus and data structures; 213 in both calculus and programming languages; 211 in both discrete mathematics and data structures; 43 in both discrete mathematics and programming languages; and no student may take calculus and discrete mathematics, or data structures and programming languages, concurrently?

19. Find the number of positive integers not exceeding 100 that are either odd or the square of an integer.

20. Suppose in a bushel of 100 apples, there are 20 that have worms in them and 15 that have bruises. Only those apples with neither worms or bruises can be sold. If there are ten bruised apples that have worms in them, how many of the 100 apples can be sold?

21. In how many ways can seven different jobs be assigned to four different employees so that each employee is assigned at least one job and the most difficult job is assigned to the best employee?

3

Graphs

3.1 Introduction

Graphs are mathematical discrete structures which have major role in computer science (algorithms and computation), electrical engineering (communication networks and coding theory), operations research (scheduling), and in many fields of engineering and also in sciences such as chemistry, biochemistry (genomics), biology, linguistics, sociology, and other fields. For instance, graphs are encoded to represent the relationship between objects. Many real-world situations can conveniently be described by means of a diagram consisting of a set of points together with lines joining certain pairs of these points. For example, the points could represent people, with lines joining pairs of friends; or the points might be communication centres, with lines representing communication links. Notice that in such diagrams, one is mainly interested in whether or not two given points are joined by a line; the manner in which they are joined is immaterial. A mathematical abstraction of situations of this type gives rise to the concept of a graph.

In this chapter, we focus on the terminology of graphs, its various types, connectivity of graphs, Eulerian path, and Hamiltonian path. Graph theory is introduced as an abstract mathematical system. The most common representation of a graph is by means of a diagram, in which the vertices are represented as points and each edge as a line segment joining its end vertices.

3.2 Graphs and Graph Models

Definition 3.2.1 Graph: *A graph $G = (V, E, \phi)$ consists of a non-empty set $V = \{v_1, v_2, \dots\}$ called the set of vertices of the graph and $E = \{e_1, e_2, \dots\}$ called the set of edges of the graph, and ϕ is a mapping from the set of edges E to the set of ordered or unordered pair of elements of vertices.*

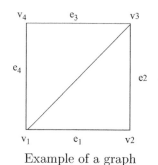

Example of a graph

Definition 3.2.2 Self-loop: *An edge having same vertex as both its end vertices is called a self-loop.*

Example 3.2.3 *Here e is a self-loop.*

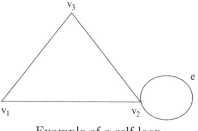

Example of a self-loop

Definition 3.2.4 Parallel Edges: *If more than one edge has the same pair of end vertices, then the edges are called parallel edges.*

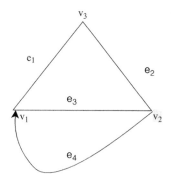

Example of parallel edges

Example 3.2.5 *In the graph in*

    (i) e_1, e_2 *are incident with v_1.*

    (ii) e_2, e_3 *are incident with v_5.*

Here, e_3 and e_4 are parallel edges.

Definition 3.2.6 *If v_i is an end vertex of some edge e_j, then e_j is said to be incident at v_i.*

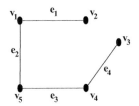

Example of incident edges on vertices

Example 3.2.7

The edge e_1 is incident at the vertex v_1. The edge e_3 is incident at the vertex v_5.

Definition 3.2.8 Adjacent Edges and Vertices: *Two non-parallel edges are said to be adjacent if they are incident on a common vertex.*

Two vertices are said to be adjacent if they are the end vertices of the same edge.

Example 3.2.9

Here, e_1 is adjacent to e_2 and e_3, and v_1 is adjacent to v_2 and v_3.

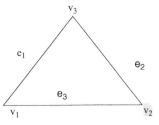

Adjacent vertices and edges

Definition 3.2.10 Simple Graph: *A graph which has neither self-loops nor parallel edges is called a simple graph as shown below.*

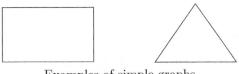

Examples of simple graphs

Definition 3.2.12 Isolated Vertex: *A vertex having no edge incident on it is called an isolated vertex as shown below.*

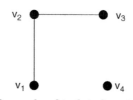

Example of isolated vertex

Definition 3.2.14 Directed Graph: *A graph in which every edge is directed is called a directed graph or digraph as shown below.*

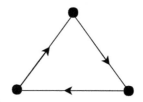

Example of directed graph

Definition 3.2.16 Undirected Graph: *A graph in which every edge is undirected is called an undirected graph.*

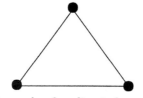

Example of undirected graph

Definition 3.2.18 Mixed Graph: *A graph in which some edges are directed and some are undirected is called a mixed graph.*

Definition 3.2.19 Multigraph: *A graph which contains some parallel edges is called a multigraph.*

Definition 3.2.20 Pseudo Graph: *A graph in which loops and parallel edges are allowed is called a pseudograph.*

3.3 Graph Terminology and Special Types of Graphs

Definition 3.3.1 Degree of a vertex: *The number of edges incident at the vertex v_i in an undirected graph is called the degree of the vertex v_i. But a loop at a vertex contributes twice to the degree of that vertex. The degree of the vertex v_i is denoted by $deg(v_i)$.*

Definition 3.3.2 In-degree and Out-degree of a vertex: *In a directed graph, the in-degree of a vertex v_i is denoted by $deg^-(v_i)$ and defined by the number of edges with v_i as their terminal vertex.*

The out-degree of a vertex v_j is denoted by $deg^+(v_j)$ and defined as the number of edges with their initial.

Theorem 3.3.3 Handshaking Theorem: *Let $G = (V, E)$ be an undirected graph with e edges. Then, $\sum_{v \in V} deg(v) = 2e$.*

Proof.
Since every edge is incident with exactly two vertices, every edge contributes two to the sum of degrees of the vertex.

Therefore, all the edges e contribute $2e$ to the sum of degrees of the vertex.

$\therefore \quad \sum_{v \in V} = 2e.$

Theorem 3.3.4 *The maximum number of edges in a simple graph with n vertices is $\dfrac{n(n-1)}{2}$.*

Proof.
From handshaking theorem, we have

$$\sum_{v \in V} deg(v) = 2e,$$

where e is the number of edges with n vertices in the graph G. That is,

$$deg(v_1) + deg(v_2) + \cdots + deg(v_n) = 2e. \qquad (3.1)$$

We know that the maximum degree of each vertex in the graph G can be $(n-1)$.

$$\therefore \quad (3.1) \implies (n-1) + (n-1) + \ldots \text{ton terms} = 2e$$
$$\implies n(n-1) = 2e$$
$$\implies e = \frac{n(n-1)}{2}.$$

Hence, the maximum number of edges in any simple graph with n vertices is $\dfrac{n(n-1)}{2}$.

Theorem 3.3.5 *A simple graph with at least two vertices has at least two vertices of same degree.*

Proof.
Let G be a simple graph with $n \geq 2$ vertices.

The graph G has no loop and parallel edges.

Hence, the degree of each vertex is $\leq n - 1$.

Suppose that all the vertices of G are of different degrees.

Following degrees $0, 1, 2, \ldots, n - 1$ are possible for n vertices of G.

Let u be the vertex with degree 0. Then, u is an isolated vertex.

Let v be the vertex with degree $n-1$. Then, v has $n-1$ adjacent vertices.

Since v is not an adjacent vertex of itself, every vertex of G other than u is an adjacent vertex of G other than u.

Hence, u cannot be an isolated vertex; this contradiction proves that a simple graph contains two vertices of same degree.

Note: The converse of the above theorem is not true.

Theorem 3.3.6 *The number of odd degree vertices is always even.*

Proof.

Let $G = (V, E)$ be any graph with n number of vertices and e number of degrees.

Let $v_1, v_2, \ldots, v_k$ be the vertices of odd degree and $v'_1, v'_2, \ldots, v'_m$ be the vertices of even degree.

To prove, k is even.

We know that $\sum_{v \in V} deg(v) = 2|E| = 2e$.

$\implies \quad \sum_{i=1}^{k} deg(v_i) + \sum_{j=1}^{m} deg(v'_j) = 2e.$

Clearly, $\sum_{j=1}^{m} deg(v'_j)$ and $2e$ are even numbers.

That is, $\sum_{i=1}^{k} deg(v_i) = $ an even number.

Since each term $deg(v_i)$ is odd, the number of terms in the left-hand side sum must be even.

$\implies \quad k$ is even.

Hence, the theorem is proved.

3.3.1   Solved Problems

1. Can a simple graph exist with 15 vertices each of degree 5?

   Solution.

   We know that the number of odd degree vertices is even.

   Hence, the number of odd degree vertices to be odd is not possible.

   We cannot say that a simple graph exists with 15 vertices each of degree 5.

2. How many vertices does a regular graph of degree 4 with ten edges have?

   Solution.

   We know that

   $$\sum_{v \in V} deg(v) = 2e$$
   $$n \times 4 = 2 \times 10$$
   $$n = \frac{20}{4}$$
   $$n = 5.$$

   $\therefore \quad$ The number of vertices is five.

3. Is there a simple graph corresponding to the following degree sequences?

    (i) $(1, 1, 2, 3)$

    (ii) $(2, 2, 4, 6)$

    Solution.

    (i) There are odd number (3) of degree vertices 1, 1, and 3. Hence, there does not exist a graph corresponding to this degree sequence.

    (ii) The number of vertices in the graph is four, and the maximum degree of a vertex is 6 which is not possible as the maximum degree cannot be one less than the number of vertices.

4. Show that in a group, there must be two people who know the same number of other people in the group.

    Solution.
    Construct the simple graph model in which V is the set of people in the group, and there is an edge associated with (u, v) if u and v know each other. Then, the degree of vertex v is the number of people v knows.

    We know that there are two vertices with the same degree. Therefore, there are two people who know the same number of other people in the group.

5. Show that the degree of a vertex of a simple graph G of n vertices cannot exceed $n - 1$.

    Solution.
    Let v be a vertex of G because G is simple and no multiple edges or loops are allowed in G. Thus, v can be adjacent to at most all the remaining $n - 1$ vertices of G. Hence, v may be of maximum degree $n - 1$ in G.

    Then, $0 \leq deg(v) \leq n - 1$, or all $v \in V$.

6. How many edges are there in a graph with ten vertices each of degree 6?

    Solution.
    Sum of the degrees of the ten vertices is

    $$(6) \times (10) = 60$$
    $$\implies 2e = 60$$
    $$\implies e = 30.$$

7. Show that the sum of degrees of all the vertices in a graph G is even.

    Solution.
    Each edge contributes two degrees in a graph.

Also, each edge contributes one degree to each of the vertices on which it is incident. Hence, if there are N edges in G, then

$$2N = deg(v_1) + deg(v_2) + \cdots + deg(v_N).$$

Thus, $2N$ is always even.

Definition 3.3.7 Regular Graph: *If every vertex of a simple graph has the same degree, then the graph is called a regular graph.*

If every vertex in a regular graph has degree k, then the graph is called k-regular graph.

Note:

1. Every null graph is regular of degree 0.

2. The complete graph K_n is of degree $n - 1$.

3. If a graph G has n vertices and is regular of degree k, then G has $\dfrac{rn}{2}$ edges.

Definition 3.3.8 Complete Graph: *A simple graph with n vertices is said to be a complete graph if the degree of every vertex is $n - 1$.*

(or)

In a graph G, if every vertex v is adjacent to all other vertices, then G is called a complete graph.

The complete graph with n vertices is denoted by K_n.

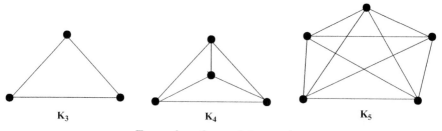

K_3 K_4 K_5

Examples of complete graphs

Example 3.3.9

Definition 3.3.10 Subgraph: *A graph $H = (V', E')$ is called a subgraph of a graph $G = (V, E)$ if $V' \subseteq V$ and $E' \subseteq E$.*

Definition 3.3.11 Cycle Graph: *A cycle graph of order n is a connected graph whose edges form a cycle of length n and is denoted by C_n.*

Example 3.3.12

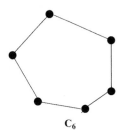

C_6

Example of cycle graph

Note:

1. In a graph, a cycle that is not a loop must have length at least three, but there may be cycles of length two in a multigraph.

2. A simple digraph having no cycles is called a cyclic graph.

3. A cyclic graph cannot have any loops.

4. The cycle C_n, $n \geq 3$, consists of n vertices $1, 2, \ldots, n$ and edges $\{1, 2\}, \{2, 3\}, \ldots, \{n-1, n\}$.

Definition 3.3.13 Wheel Graph: *A wheel graph of order n is obtained by joining a new vertex called "Hub" to each vertex of a cycle graph of order $n - 1$, denoted by W_n.*

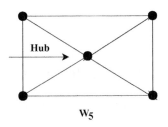

W_5

Example of wheel graph

Example 3.3.14

Note: We obtain the wheel W_n when we add an additional vertex to the cycle C_n, for $n \geq 3$, and connect this new vertex to each of the n vertices in C_n, by new edges.

Definition 3.3.15 Bipartite Graph: *A graph G is said to be bipartite if its vertex set V can be partitioned into two disjoint non-empty sets V_1 and V_2 such that $V_1 \cup V_2$ and every edge in E has one end in V_1 and the other end in V_2.*

Example 3.3.16 *Let $V = V_1 \cup V_2$ where*
   $V_1 = \{u_1, u_3, u_5, u_7\}$ and $V_2 = \{u_2, u_4, u_6, u_8\}$.
Then, G is a bipartite graph shown below.

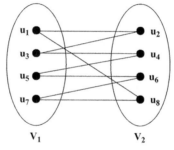

Example of bipartite graph

Definition 3.3.17 Complete Bipartite Graph: *A bipartite graph G with the bipartition V_1 and V_2 is called complete bipartite if every vertex in V_1 is adjacent to every vertex in V_2.*

A complete bipartite graph that may be partitioned into sets A and B as above such that $|A| = a$ and $|B| = b$ is denoted by $K_{a,b}$.

Example 3.3.18 *The graph $K_{3,3}$ is a complete bipartite graph.*

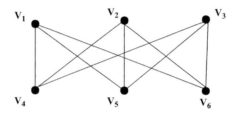

$$K_{3,3}$$

Example of complete bipartite graph

Definition 3.3.19 Star Graph: *Any graph that is $K_{1,n}$ is called a star graph.*

Example 3.3.20 *The graph $K_{1,6}$ is a star graph.*

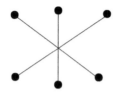

$$K_{1,6}$$

Example of star graph

3.3.2   Graph Colouring

The assignment of colours to the vertices of G, one colour to each vertex, so that adjacent vertices are assigned different colours is called the proper colouring of G or simply vertex colouring.

If G has n colouring, then G is said to be n-colourable.

Theorem 3.3.21 *A simple graph is bipartite if and only if it is possible to assign one of two different colours to each vertex of the graph so that no two adjacent vertices are assigned the same colour.*

Proof.

Let $G = (V, E)$ be a bipartite simple graph. Then, $V = V_1 \cup V_2$, where V_1 and V_2 are disjoint sets and every edge in E connects a vertex in V_1 and a vertex in V_2.

If we assign one colour to each vertex in V_1 and a second colour to each vertex in V_2, then no two adjacent vertices are assigned the same colour.

Suppose that it is possible to assign colours to the vertices of the graph using just two colours.

$\implies$ No two adjacent vertices are assigned the same colour.

Let V_1 be the set of vertices assigned one colour and V_2 be the set of vertices assigned the other colour. Then, V_1 and V_2 are disjoint and $V = V_1 \cup V_2$.

That is, every edge connects a vertex in V_1 and a vertex in V_2 since no two adjacent vertices are either both in V_1 or both in V_2. Consequently, G is bipartite.

3.3.3   Solved Problems

1. What is the degree sequence of K_n, where n is positive integer? Explain your answer.

    Solution.

    Each of the n vertices is adjacent to each of the other $n - 1$ vertices, so the degree sequence is $n - 1, n - 1, \ldots, n - 1$ (n terms).

2. Determine whether each of the following sequences is a graph. For those that are, draw a graph having the given degree sequence.

    (i) 5, 4, 3, 2, 1

    (ii) 3, 2, 2, 1, 0

    (iii) 1, 1, 1, 1, 1

    Solution.

    (i) No, since the sum of degrees $= 5 + 4 + 3 + 2 + 1 = 15$ which is odd.

    (ii) Yes.

    (iii) No, since the sum of degrees $= 1 + 1 + 1 + 1 + 1 = 5$ which is odd.

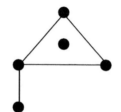

Graph of the given sequence

3. How many vertices and edges are there in K_n?

   Solution.

   K_n has n vertices and $\dfrac{n(n-1)}{2}$ edges.

4. Find the degree sequence of each of the following graphs.

   (i) K_4

   (ii) K_5

   (iii) K_2

   Solution.

   (i) 3, 3, 3, 3

   (ii) 4, 4, 4, 4, 4

   (iii) 1, 1

5. How many vertices and edges do the following graphs have?

   (i) C_n

   (ii) C_8

   (iii) Also find the degree sequence of C_4.

   Solution.

   (i) n vertices and n edges

   (ii) Eight vertices and eight edges

   (iii) 2, 2, 2, 2.

6. Show that C_6 is a bipartite graph?

   Solution.
   The vertex set of C_6 can be partitioned into the two sets $V_1 = \{v_1, v_3, v_5\}$ and $V_2 = \{v_2, v_4, v_6\}$, and every edge of C_6 connects a vertex in V_1 and a vertex in V_2. Hence, C_6 is a bipartite graph.

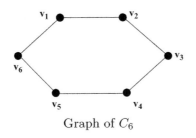

Graph of C_6

7. Is K_3 bipartite?

   Solution.

   No, the complete graph K_3 is not bipartite as shown below.

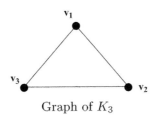

Graph of K_3

   If we divide the vertex set of K_3 into two disjoint sets, one of the two sets must contain two vertices. If the graph is bipartite, these two vertices should not be connected by an edge, but in K_3 each vertex is connected to every other vertex by an edge.

   $\therefore$ K_3 is not bipartite.

8. How many vertices and edges are there in a complete bipartite graph $K_{m,n}$?

   Solution.

   There are $m + n$ vertices and mn edges.

9. Find the degree sequence of the graph $K_{2,3}$.

   Solution.

   3, 3, 2, 2, 2.

10. For which values of m and n is $K_{m,n}$ regular?

    Solution.

    A complete bipartite graph $K_{m,n}$ is not regular if $m \neq n$.

    $\implies$ If $m = n$, then $K_{m,n}$ is regular.

11. Prove that a graph which contains a triangle cannot be bipartite.

    Solution.

    At least two of the three vertices must lie in one of the bipartite sets because these two are joined by edge; thus, the graph cannot be bipartite.

12. Show that if G is a bipartite simple graph with v vertices and e edges, then $e \leq \dfrac{v^2}{4}$.

Solution.

Let G be a complete bipartite graph with v vertices.

Let v_1 and v_2 be the number of vertices in the partitions V_1 and V_2 of vertex set of G.

Since G is complete bipartite, each vertex in V_1 is joined to each vertex in V_2 by exactly one edge.

Thus, G has $v_1 v_2$ edges when $v_1 + v_2 = v$.

But we know the maximum value of $v_1 v_2$ subject to $v_1 + v_2 = v$ is $\dfrac{v^2}{4}$.

Thus, the maximum number of edges in G is $\dfrac{v^2}{4}$.

That is, $e \leq \dfrac{v^2}{4}$.

13. Show that the graph G is bipartite.

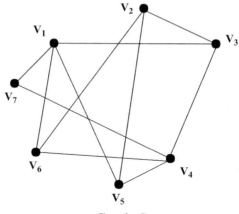

Graph G

Solution.

Graph G is bipartite since its vertex set is the union of two disjoint sets $\{v_1, v_2, v_3\}$ and $\{v_4, v_5, v_6, v_7\}$ and each edge connects a vertex in one of these subsets to a vertex in the other subset.

14. Draw the complete bipartite graphs $K_{2,3}$, $K_{3,3}$, $K_{3,5}$, and $K_{2,6}$.

Solution.

The complete bipartite graphs are shown below.

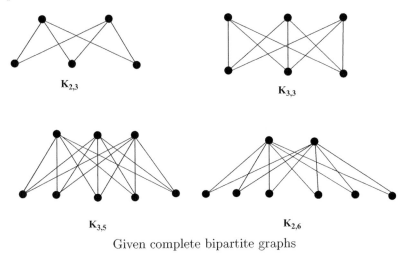

K_{2,3}

K_{3,3}

K_{3,5}

K_{2,6}

Given complete bipartite graphs

15. How many subgraphs with at least one vertex does K_3 have?

Solution. 17

3.4   Representing Graphs and Graph Isomorphism

We can represent a simple graph in the form of edge list or in the form of adjacency lists which may be useful in computer programming.

Definition 3.4.1 Adjacency matrix of a simple graph: *Let $G = (V, E)$ be a simple graph with n vertices $\{v_1, v_2, \ldots, v_n\}$. Its adjacency matrix is denoted by $A = (a_{ij})$ and defined by*

$$a_{ij} = \begin{cases} 1, & \text{if } v_i \text{ and } v_j \text{ are adjacent} \\ 0, & \text{otherwise.} \end{cases}$$

Example 3.4.2

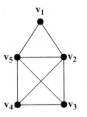

Example of a graph with adjacency matrix

The adjoint matrix for the graph in the figure above is

$$A = \begin{array}{c} \\ v_1 \\ v_2 \\ v_3 \\ v_4 \\ v_5 \end{array} \begin{array}{ccccc} v_1 & v_2 & v_3 & v_4 & v_5 \\ \left[\begin{array}{ccccc} 0 & 1 & 0 & 0 & 1 \\ 1 & 0 & 1 & 1 & 1 \\ 0 & 1 & 0 & 1 & 1 \\ 0 & 1 & 1 & 0 & 1 \\ 1 & 1 & 1 & 1 & 0 \end{array} \right] \end{array}$$

Definition 3.4.3 Incidence Matrix: *Let $G = (V, E)$ be a graph with n vertices $v_1, v_2, \ldots, v_n$ and m edges $e_1, e_2, \ldots, e_m$. Then, the $n \times m$ matrix $B = (b_{ij})$ where*

$$b_{ij} = \begin{cases} 1, & \text{if the edge } e_j \text{ is incident on } v_i \\ 0, & \text{otherwise.} \end{cases}$$

Example 3.4.4

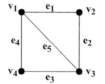

Example of a graph with incidence matrix

The incidence matrix for the graph in the figure above is

$$B = \begin{array}{c} \\ v_1 \\ v_2 \\ v_3 \\ v_4 \end{array} \begin{array}{ccccc} e_1 & e_2 & e_3 & e_4 & e_5 \\ \left[\begin{array}{ccccc} 1 & 0 & 0 & 1 & 1 \\ 1 & 1 & 0 & 0 & 0 \\ 0 & 1 & 1 & 0 & 1 \\ 0 & 0 & 1 & 1 & 0 \end{array} \right] \end{array}$$

Observations about the incidence matrix:

1. Since every edge is incident on exactly two vertices, each column of B has exactly two 1's.

2. The number of 1's in each row is equal to the degree of the corresponding vertex.

3. A row with all 0's represents an isolated vertex.

4. Parallel edges in a graph produce identical columns in its incidence matrix.

Definition 3.4.5 Isomorphic Graphs: *The simple graphs $G_1 = (V_1, E_1)$ and $G_2 = (V_2, E_2)$ are isomorphic if there is a one-to-one and onto function f from V_1 to V_2 with the property that a and b are adjacent in G_1 if and only if $f(a)$ and $f(b)$ are adjacent in G_2, for all a and b in V_1. Such a function f is called an isomorphism.*

In other words, two graphs G_1 and G_2 are isomorphic if there is a function $f : V(G_1) \longrightarrow V(G_2)$ from the vertices of G_1 to the vertices of G_2 such that

(i) *f is one-to-one*
(ii) *f is onto and*
(iii) *for each pair of vertices u and v of G_1*

$$[u, v] \in E(G_1) \Leftrightarrow [f(u), f(v)] \in E(G_2).$$

Any function f with the above three properties is called an isomorphism from G_1 to G_2.

Example 3.4.6 *Consider the graphs G_1 and G_2 in the following figure. Let $f : G_1 \longrightarrow G_2$ be a function with $f(u_1) = v_1$, $f(u_2) = v_4$, $f(u_3) = v_3$, $f(u_4) = v_2$. Then, f is a one-to-one and onto function between G_1 and G_2. Here, f preserves the adjacency. The adjacent vertices in G_1 are u_1 and u_2, u_1 and u_3, u_2 and u_4, and u_3 and u_4, and each of the pairs $f(u_1) = v_1$ and $f(u_2) = v_4$, $f(u_1) = v_1$ and $f(u_3) = v_3$, $f(u_2) = v_4$ and $f(u_4) = v_2$, and $f(u_3) = v_3$ and $f(u_4) = v_2$ are adjacent in G_2. Hence, the graphs G_1 and G_2 are isomorphic.*

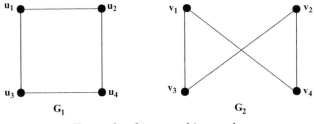

Example of isomorphic graphs

3.4.1 Solved Problems

1. Write the adjacency matrix of the following graph.

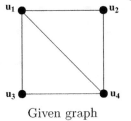

Given graph

Solution.

The adjacency matrix is

$$
A = \begin{array}{c} \\ u_1 \\ u_2 \\ u_3 \\ u_4 \end{array}
\begin{array}{cccc} u_1 & u_2 & u_3 & u_4 \\ \left[\begin{array}{cccc} 0 & 1 & 1 & 1 \\ 1 & 0 & 0 & 1 \\ 1 & 0 & 0 & 1 \\ 1 & 1 & 1 & 0 \end{array}\right] \end{array}
$$

2. Write the incidence matrix for the following graph.

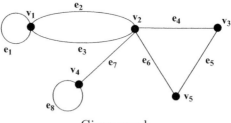

Given graph

Solution.

The incidence matrix is

$$
B = \begin{array}{c} \\ v_1 \\ v_2 \\ v_3 \\ v_4 \\ v_5 \end{array}
\begin{array}{cccccccc} e_1 & e_2 & e_3 & e_4 & e_5 & e_6 & e_7 & e_8 \\ \left[\begin{array}{cccccccc} 1 & 1 & 1 & 0 & 0 & 0 & 0 & 0 \\ 0 & 1 & 1 & 1 & 0 & 1 & 1 & 0 \\ 0 & 0 & 0 & 1 & 1 & 0 & 0 & 0 \\ 0 & 0 & 0 & 0 & 0 & 0 & 0 & 0 \\ 0 & 0 & 0 & 0 & 1 & 1 & 0 & 0 \end{array}\right] \end{array}
$$

3. What is the sum of the entries in a row of the incidence matrix for an undirected graph?

Solution.

Sum is 2 if the edge e is not a loop and 1 if the edge e is a loop.

4. Check whether the two graphs are isomorphic or not.

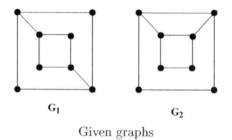

Given graphs

Solution.

In the graph G_1, vertices of degree 2 are not adjacent, while in the graph G_2, vertices of degree 2 are adjacent. Since isomorphism preserves adjacency of vertices, the graphs are not isomorphic.

5. Prove that the graphs G_1 and G_2 are isomorphic.

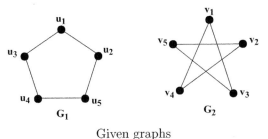

Given graphs

Solution.

The two graphs have the same number of vertices, same number of edges, and same degree sequence. Consider the function f defined by

$$f(u_1) = v_1, \quad f(u_2) = v_3, \quad f(u_3) = v_4, \quad f(u_4) = v_2, \quad f(u_5) = v_5.$$

Then, the adjacency matrices of the two graphs corresponding to f are

$$A(G_1) = \begin{array}{c} \\ u_1 \\ u_2 \\ u_3 \\ u_4 \\ u_5 \end{array} \begin{array}{ccccc} u_1 & u_2 & u_3 & u_4 & u_5 \\ \left[\begin{array}{ccccc} 0 & 1 & 1 & 0 & 0 \\ 1 & 0 & 0 & 0 & 1 \\ 1 & 0 & 0 & 1 & 0 \\ 0 & 0 & 1 & 0 & 1 \\ 0 & 1 & 0 & 1 & 0 \end{array} \right] \end{array}$$

$$A(G_2) = \begin{array}{c} \\ v_1 \\ v_2 \\ v_3 \\ v_4 \\ v_5 \end{array} \begin{array}{ccccc} v_1 & v_2 & v_3 & v_4 & v_5 \\ \left[\begin{array}{ccccc} 0 & 1 & 1 & 0 & 0 \\ 1 & 0 & 0 & 0 & 1 \\ 1 & 0 & 0 & 1 & 0 \\ 0 & 0 & 1 & 0 & 1 \\ 0 & 1 & 0 & 1 & 0 \end{array} \right] \end{array}$$

Therefore, $A(G_1) = A(G_2)$. Hence, G_1 and G_2 are isomorphic to each other.

6. Prove that any two simple connected graphs with n vertices all of degree 2 are isomorphic.

Solution.

We know that the total degree of a graph is given by

$$\sum_{i=1}^{n} deg(v_i) = 2|E|.$$

Then, $|V|$ = number of vertices = n
$|E|$ = number of edges.
Further, the degree of each vertex is 2.
$\therefore$ $\sum_{i=1}^{n} 2 = 2|E|$
$\implies$ $n = |E|$
$\therefore$ Number of edges = number of vertices. Hence, the graphs are cycle graphs. Therefore, they are isomorphic.

7. Can a simple graph with seven vertices be isomorphic to its complement?

   Solution.
   A graph with seven vertices can have a maximum number of edges
   $$= \frac{7(7-1)}{2} = \frac{7 \times 6}{2} = 21 \text{ edges.}$$

   But 21 edges cannot be splitted into two equal integers. Therefore, a graph and its complement cannot have equal number of edges. Hence, a graph with seven vertices cannot be isomorphic to its complement.

8. Let G be a simple graph, all of whose vertices have degree 3 and $|E| = 2|V| - 3$. What can be said about G?

   Solution.
   $$\sum_{i=1}^{|V|} deg(v_i) = 2|E|$$
   $$3(|V| - 1 + 1) = 2|E|$$
   $$3|V| = 2|E| = 2(2|V| - 3) = 4|V| - 6$$
   $$\implies \quad |V| = 6.$$

   The number of vertices in G is six. Hence, it can be concluded that G is isomorphic to $K_{3,3}$.

9. Show that isomorphism of simple graphs is an equivalence relation.

   Solution.

   (i) **Reflexive:** G is isomorphic to itself by the identity function. Hence, isomorphism is reflexive.

   (ii) **Symmetric:** Suppose that G is isomorphic to H. Then, there exists a one-to-one correspondence f from G to H that preserves adjacency and non-adjacency. From this, f^{-1} is a one-to-one correspondence from H to G that preserves adjacency and non-adjacency. Hence, isomorphism is symmetric.

   (iii) **Transitive:** If G is isomorphic to H and H is isomorphic to K, then there are one-to-one correspondences f and g from G to H and from H to K that preserve adjacency and non-adjacency. It follows that $g \circ f$ is a one-to-one correspondence from G to K that preserves adjacency and non-adjacency. Hence, isomorphism is transitive.

From the above (i)–(iii), isomorphism is an equivalence relation.

3.4.2 Problems for Practice

1. Write the adjacency matrix of the following graph.

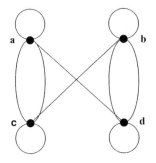

2. Draw the directed graph for the following adjacency matrix.

$$\begin{pmatrix} 0 & 0 & 1 & 1 \\ 0 & 0 & 1 & 0 \\ 1 & 1 & 0 & 1 \\ 1 & 1 & 1 & 0 \end{pmatrix}$$

3. Draw an undirected graph for the following adjacency matrix.

$$\begin{pmatrix} 1 & 2 & 0 & 1 \\ 2 & 0 & 3 & 0 \\ 0 & 3 & 1 & 1 \\ 1 & 0 & 1 & 0 \end{pmatrix}$$

4. Find the adjacency matrix of the given directed graph.

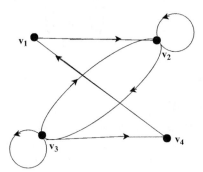

5. Check whether the graphs with the following adjacency matrices are isomorphic.

$$\begin{pmatrix} 0 & 0 & 1 \\ 0 & 0 & 1 \\ 1 & 1 & 0 \end{pmatrix}, \begin{pmatrix} 0 & 1 & 1 \\ 1 & 0 & 0 \\ 1 & 0 & 0 \end{pmatrix}$$

6. Determine whether the graphs with the following adjacency matrices are isomorphic.

$$\begin{pmatrix} 0 & 1 & 1 & 0 \\ 1 & 0 & 0 & 1 \\ 1 & 0 & 0 & 1 \\ 0 & 1 & 1 & 0 \end{pmatrix}, \begin{pmatrix} 0 & 1 & 0 & 1 \\ 1 & 0 & 0 & 0 \\ 0 & 0 & 0 & 1 \\ 1 & 0 & 1 & 0 \end{pmatrix}$$

7. Find whether the following graphs are isomorphic to each other.

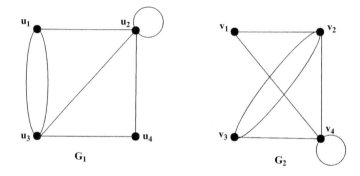

8. Find a pair of non-isomorphic graphs with the same degree sequence such that one graph is bipartite but the other graph is not bipartite.

9. What is the product of the incidence matrix and its transpose for an undirected graph?

10. Draw the graph represented by the following adjacency matrix.

$$\begin{pmatrix} 1 & 2 & 1 \\ 2 & 0 & 0 \\ 0 & 2 & 2 \end{pmatrix}$$

3.5   Connectivity

Definition 3.5.1 Walk: *A walk is defined as a finite alternating sequence of vertices and edges, beginning and ending with vertices such that each edge is incident with the vertex preceding and following it. (No edge appears more than once, and vertex may be repeated.)*

Example 3.5.2 *Consider the graph G. In this graph,* $v_1e_1v_4e_2v_3e_6v_5$ *is a walk.*

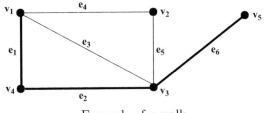

Example of a walk

Remarks:

1. A walk is also referred to as an edge train or a chain.

2. No edge appears more than once in a walk.

3. Every walk is a subgraph of G.

Definition 3.5.3 Terminal Vertex: *In a walk, the vertex that begins and ends the walk is called its terminal vertex.*

For example, in the walk $v_1e_3v_3e_5v_2$ *in the figure above, the terminal vertices are* v_1 *and* v_2.

Definition 3.5.4 Closed Walk: *A walk with same end vertices is called a closed walk.*

In the example above, $v_1e_3v_3e_5v_2e_4v_1$ *is a closed walk.*

Definition 3.5.5 Open Walk: *A walk which is not closed is called an open walk.*

In the graph above, $v_1e_3v_3e_5v_2$ *is an open walk.*

Definition 3.5.6 Path or Simple Path or Elementary Path: *An open walk in which no vertex appears more than once is called a path or simple path or an elementary path.*

For example, in the graph above, $v_1e_3v_3e_6v_5$ *is a path, but* $v_1e_3v_3e_5v_2e_4v_1e_1v_4$ *is not a path, since* v_1 *is repeated twice.*

Definition 3.5.7 Circuit or Cycle or Circular Path or Elementary Cycle: *A closed walk in which no vertex appears more than once is called a circuit or a cycle or a circular path or an elementary cycle.*

In the graph above, $v_1e_3v_3e_5v_2e_4v_1$ *is a circuit.*

Remarks:

1. A path does not intersect itself.

2. A self-loop can be included in a walk but not a path.

3. Every circuit is called as a non-intersecting walk.

4. Every self-loop is a circuit, but every circuit is not a self-loop.

5. Every vertex in a circuit is of degree 2.

3.5.1   Connected and Disconnected Graphs

Definition 3.5.8 Connected Graph: *A graph G is a connected graph if there is at least one path between every pair of vertices in G. Otherwise G is a disconnected graph.*

Example 3.5.9 *Consider the connected graph below.*

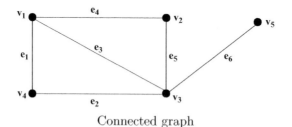

Connected graph

Definition 3.5.10 Component or Block: *A disconnected graph consists of two or more connected graphs. Each of these connected subgraphs is called a component or block.*

Example 3.5.11

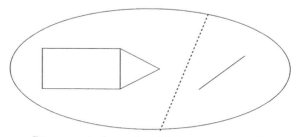

Disconnected graph with two components

Theorem 3.5.12 *A graph G is disconnected iff its vertex set V can be partitioned into two non-empty disjoint subsets V_1 and V_2 such that there exists no edge in G whose one end vertex is in the subset V_1 and the other in the subset V_2.*

Proof.

Let G be a disconnected graph.

Let us fix a vertex a in G. Let V_1 be the set of all vertices such that they are joined by paths to a.

Now, consider $V_2 = V - V_1$. Clearly, V_2 is not empty (since G is disconnected, V_1 does not contain all vertices of G). Also, no vertex in V_1 is joined to any vertex in V_2 by an edge.

Conversely, suppose the partition exists.

Consider any two arbitrary vertices a and b of G such that $a \in V_1$ and $b \in V_2$.

$\implies$ There exists no path between a and b.

Otherwise, there would be at least one edge whose one end vertex would be in V_1 and other end in V_2.

$\implies$ G is not connected.

Hence, the theorem is proved.

Theorem 3.5.13 *If a graph (connected or disconnected) has exactly two vertices of odd degree, then there must be a path joining these two vertices.*

Proof.

Let G be a graph with exactly two odd degree vertices. Let them be v_1 and v_2.

Case (i): Let G be a connected graph.

$\implies$ There exists at least one path between v_1 and v_2.

Case (ii): Let G be a disconnected graph.

$\implies$ There exist at least two components.

Without loss of generality, we can assume G contains two components g_1 and g_2.

If both v_1 and v_2 lie in the same component, then there exists a path joining between v_1 and v_2.

Suppose v_1 lies in g_1 and v_2 lies in g_2. Then, g_1 is a graph containing exactly one odd degree vertex, and g_2 is a graph having exactly one odd degree vertex, which is a contradiction to the theorem that "the number of odd degree vertices is always even".

Therefore, v_1 and v_2 must be in the same component.

Hence, the theorem is proved.

Theorem 3.5.14 *A simple graph with n vertices and k components can have at most $\dfrac{(n-k)(n-k+1)}{2}$ edges.*

Proof.

Let G be a simple graph with n vertices and k components, namely $g_1, g_2, \ldots, g_k$.

Let n_i be the number of vertices in the ith components g_i, for all $i = 1, 2, \ldots, k$.

Clearly, $n_1 + n_2 + \cdots + n_k = n$

$\implies \sum_{i=1}^{k} n_i = n$, and $n_i \geq 1$.

We know that the maximum number of edges in the ith component is $\dfrac{n_i(n_i - 1)}{2}$.

$\therefore$ The maximum number of edges in G

$=$ the number of edges in g_1 + the number of edges in g_2

$+ \cdots +$ the number of edges in g_k

$= n_1(n_1 - 1)/2 + n_2(n_2 - 1)/2 + \cdots + n_k(n_k - 1)/2$

$$= \sum_{i=1}^{k} n_i(n_i - 1)/2 = \sum_{i=1}^{k} \left(n_i^2 - n_i \right)/2 = \left(\sum_{i=1}^{k} n_i^2 - \sum_{i=1}^{k} n_i \right)/2$$

$$= \left(\sum_{i=1}^{k} n_i^2 - n \right)/2. \tag{3.2}$$

Let us consider

$$\sum_{i=1}^{k} (n_i - 1) = \sum_{i=1}^{k} n_i - \sum_{i=1}^{k} 1$$

$$= (n - k).$$

Squaring on both sides, we get

$$\left[\sum_{i=1}^{k} (n_i - 1) \right]^2 = (n - k)^2$$

(or) $[(n_1 - 1) + (n_2 - 1) + \cdots + (n_k - 1)]^2 = (n - k)^2$

$[(n_1 - 1)^2 + (n_2 - 1)^2 + \cdots + (n_k - 1)^2]$

$+ \text{(positive cross terms)} = (n - k)^2$

(or) $[(n_1^2 + 1 - 2n_1) + (n_2^2 + 1 - 2n_2) + \cdots + (n_k^2 + 1 - 2n_k)]$

$+ \text{(positive cross terms)} = (n - k)^2$

(or) $(n_1^2 + n_2^2 + \cdots + n_k^2) + \underbrace{(1 + 1 + \cdots + 1)}_{k \ times}$

$- 2(n_1 + n_2 + \cdots + n_k) + \text{(positive cross terms)} = (n - k)^2$

(or) $\sum_{i=1}^{k} n_i^2 - k - 2 \sum_{i=1}^{k} n_i + \text{(positive cross terms)} = (n - k)^2$

(or) $\sum_{i=1}^{k} n_i^2 + k - 2 \sum_{i=1}^{k} n_i \leq (n - k)^2 \text{(by omitting the positive cross terms)}$

(or) $\sum_{i=1}^{k} n_i^2 + k - 2n \leq (n - k)^2$

(or) $\sum_{i=1}^{k} n_i^2 \leq (n - k)^2 - k + 2n. \tag{3.3}$

Therefore, the maximum number of edges in G

$$= \frac{1}{2} \sum_{i=1}^{k} n_i^2 - \frac{n}{2} \quad \text{[using (3.2)]}$$
$$\leq [(n-k)^2 - k + 2n - n]/2 \quad \text{[using (3.3)]}$$
$$= [(n-k)^2 - k + n]/2$$
$$= (n-k)(n-k+1)/2.$$

Hence, the theorem is proved.

3.6 Eulerian and Hamiltonian Paths

Definition 3.6.1 Eulerian Circuit: *An Eulerian circuit in a graph G is a simple circuit containing every edge of G.*

Definition 3.6.2 Eulerian Trail: *A trail in G is called an Eulerian trail if it includes each edge of G exactly once.*

Definition 3.6.3 Eulerian Path: *An Eulerian path in G is a simple path containing every edge of G.*

Definition 3.6.4 Eulerian Graph: *A closed walk which contains all edges of the graph G is called an Euler line, and the graph containing at least one Euler line is called an Eulerian graph.*

Example 3.6.5 *The graphs are Eulerian graphs.*

Star of David   Mohammed's Semi Stars
Examples of Eulerian graphs

Theorem 3.6.6 *A given connected graph G is an Eulerian graph if and only if all vertices of G are of even degree.*

Proof.

Suppose G is an Eulerian graph.

$\implies$ G contains an Euler line.

$\implies$ G contains a closed walk covering all edges.

To prove: All vertices of G are of even degree.

In tracing the closed walk, every time the walk meets a vertex v, it goes through two new edges incident on v with one we "entered" and other "exited". Since it is a closed walk, this is true for all vertices (intermediate and terminal vertices). Thus, the degree of every vertex is even.

Conversely, suppose that all vertices of G are of even degree.

To prove: G is an Eulerian graph.

That is, to prove G contains an Euler line.

Construct a closed walk starting at an arbitrary vertex v and going through the edges of G such that no edge is repeated. Since each vertex is of even degree, we can exit from each and every vertex where we enter, and the tracing can stop only at the vertex v. Name the closed walk as h.

Case (i): If h covers all edges of G, then h becomes an Euler line, and hence G is an Eulerian graph.

Case (ii): If h does not cover all edges of G, then remove all edges of h from G and obtain the graph G'. Since both G and G' have vertices which are of even degree, every vertex in G' is also of even degree.

Since G is connected, h will touch G' in at least one vertex v'. Starting from v', we can again construct a new walk h' in G'. This will terminate only at v', since every vertex in G' is also of even degree.

Now, this walk h' combined with h forms a closed walk, starts and ends at v, and has more edges than h.

This process is repeated until we obtain a closed walk covering all edges of G. Thus, G is an Eulerian graph.

Theorem 3.6.7 *A connected multigraph with at least two vertices has an Eulerian circuit if and only if each of its vertices has even degree.*

Proof.

Given: A connected multigraph G has an Eulerian path but not an Eulerian circuit.

To prove: G has exactly two vertices of odd degree.

Suppose that a connected multigraph has an Eulerian path from a to b but not an Eulerian circuit. The first edge of the path contributes 1 to the degree of a. A contribution of 2 to the degree of a is made every time the path passes through a. The last edge in the path contributes 1 to the degree of b.

Every time the path goes through b, there is a contribution of 2 to its degree. Consequently, both a and b have odd degree. Every other vertex has even degree, because the path contributes 2 to the degree of a vertex whenever it passes through it.

Theorem 3.6.8 *A connected multigraph has an Eulerian path but not an Eulerian circuit if and only if it has exactly two vertices of odd degree.*

Proof.

Given: The graph has exactly two vertices of odd degree.

To prove: G has an Eulerian path.

Suppose that a graph has exactly two vertices of odd degree, say a and b. Consider the larger graph made up of the original graph with the addition of an edge $\{a, b\}$.

Every vertex of this larger graph has even degree, so there is an Eulerian circuit. The removal of the new edge produces an Eulerian path in the original graph.

Chinese postman problem:

If a postman can find an Eulerian path in the graph that represents the streets the postman needs to cover, this path produces a route that traverses each street of the route exactly once. If no Eulerian path exists, some streets will have to be traversed more than once. This problem is known as the Chinese postman problem.

Theorem 3.6.9 *A directed multigraph having no isolated vertices has an Eulerian path but not an Eulerian circuit if and only if the graph is weakly connected and the in-degree and out-degree of each vertex are equal for all but two vertices, one that has in-degree larger than its out-degree by 1 and the other that has out-degree larger than its in-degree by 1.*

Proof.

If there is an Eulerian path, as we follow, each vertex except the starting and ending vertices must have equal in-degree and out-degree, since whenever we come to a vertex along an edge, we leave it along another edge. The starting vertex must have out-degree 1 larger than its in-degree, since we use one edge leading out of this vertex and whenever we visit it again, we use one edge leading into it and one leaving it.

Similarly, the ending vertex must have in-degree 1 greater than its out-degree. Since the Eulerian path with directions erased produces a path between any two vertices, in the underlying undirected graph, the graph is weakly connected.

Conversely, suppose the graph meets the degree conditions stated. If we add one more edge from the vertex of deficient out-degree to the vertex X with equal in-degree and out-degree. Because the graph is still weakly connected, by this new graph has an Eulerian circuit. Now, delete the added edge to obtain the Eulerian path.

3.6.1 Hamiltonian Path and Hamiltonian Circuits

Definition 3.6.10 Hamiltonian Path: *A simple path in a graph G that passes through every vertex exactly once is called a Hamiltonian path. That is, the simple path $x_0, x_1, x_2, \ldots, x_{n-1}, x_n$ in the graph $G = (V, E)$ is a Hamiltonian path if $V = \{x_0, x_1, x_2, \ldots, x_{n-1}, x_n\}$ and $x_i \neq x_j$ for $0 \leq i < j \leq n$.*

Definition 3.6.11 *A simple circuit in a graph G that passes through every vertex exactly once is called a Hamiltonian circuit. And the simple* *circuit* $x_0, x_1, \ldots, x_{n-1}, x_n, x_0$ *(with* $n > 0$*) is a Hamiltonian circuit if* $x_0, x_1, \ldots, x_{n-1}, x_n$ *is a Hamiltonian path.*

Definition 3.6.12 Unicursal Graph: *An open walk that includes all edges of G without retracing any edge is called a unicursal line or an open Euler line.*

A connected graph which has a unicursal line is called a unicursal graph.

Example 3.6.13 In the graph, the unicursal line is $v_2 e_1 v_4 e_2 v_3 e_3 v_2 e_4 v_1$.

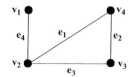

Example of unicursal graph

Remark:
1. Adding an edge between the initial and final vertices of unicursal line, we obtain an Euler line.
2. A connected graph is unicursal, if it has exactly two odd degree vertices.

Theorem 3.6.14 *In a connected graph G, with exactly $2k$ odd degree vertices, there exist k edge-disjoint subgraphs such that they together contain all edges of G and that each is a unicursal graph.*

Proof.
Let the odd degree vertices of the given graph be named as $v_1, v_2, \ldots, v_k$ and $\omega_1, \omega_2, \ldots, \omega_k$ in any arbitrary order.

Add k edges (new edges) to G between the pair of vertices (v_1, ω_1), $(v_2, \omega_2), \ldots, (v_k, \omega_k)$ to form a new graph G'.

In the resultant graph G', every vertex is of even degree.

$\implies$ G' is an Eulerian graph.

$\implies$ G' contains an Euler line, say P.

If we remove k newly added edges from P, that will split P into k walks each of a unicursal line to itself.

That is, the first removal will leave a single unicursal line. The second removal will split that into two unicursal lines, and each successive removal will split a unicursal line into two unicursal lines, until there are k of them.

Hence, the theorem is proved.

3.6.2 Solved Problems

1. Does the graph given below have a Hamiltonian path? If so, find such a path. If it does not, give an argument to show why no such path exists.

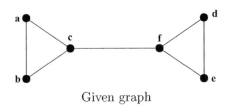

Given graph

Solution.

$a–b–c–f–d–e$ is a Hamiltonian path.

2. Does the graph below have a Hamiltonian path. If so, find such a path. If it does not, give an argument to show why no such path exists.

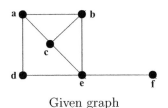

Given graph

Solution.

$f–e–d–a–b–c$ is a Hamiltonian path.

3. Does the graph given below have a Hamiltonian path? If so, find such a path. If it does not, give an argument to show why no such path exists.

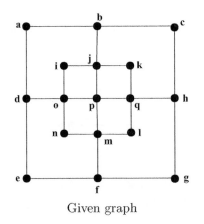

Given graph

Solution.

No Hamiltonian path exists. There are eight vertices of degree 2, and only two of them can be end vertices of a path. For each of the other six, their two incident edges must be in the path. It is easy to see that if there is to be a Hamiltonian path, exactly one of the inside corner vertices must be an end and that this is impossible.

4. Show that K_n has a Hamiltonian circuit whenever $n \geq 3$.

   Solution.

   We can form a Hamiltonian circuit in K_n beginning at any vertex. Such a circuit can be built by visiting vertices in any order we choose, as long as the path begins and ends at the same vertex and visits each of the other vertices exactly once.

   It is possible since there are edges in K_n between any two vertices.

5. Give an example of a graph that has an Eulerian circuit and a Hamiltonian circuit, which are distinct.

   Solution.

   The graph having an Eulerian circuit and a Hamiltonian circuit which are distinct as shown below. The Eulerian circuit is a–c–b–c–d–b–a.

   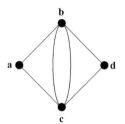

   Graph with Eulerian and Hamiltonian circuits

   The Hamiltonian circuit is a–b–d–c–a.

6. Give an example of a graph which has a Hamiltonian circuit but not an Eulerian circuit.

   Solution.

   The graph having a Hamiltonian circuit but not an Eulerian circuit as shown below.

   Graph with Hamiltonian circuit but no Eulerian circuit

   The Hamiltonian circuit is a–b–d–c–a.

   There is no Eulerian circuit.

7. Give an example of a graph which has an Eulerian circuit but not a Hamiltonian circuit.

Solution.

The graph having an Eulerian circuit but not a Hamiltonian circuit as shown below.

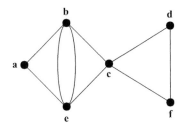

Graph with Eulerian circuit but no Hamiltonian circuit

The Eulerian circuit is a–e–b–e–c–d–f–c–b–a.

There is no Hamiltonian circuit.

8. Show that a bipartite graph with an odd number of vertices does not have a Hamiltonian circuit.

Solution.
Suppose that $G = (V, E)$ is a bipartite graph with $V = V_1 \cup V_2$, where no edge connects a vertex in V_1 and a vertex in V_2.

Suppose that G has a Hamiltonian circuit. Such a circuit must be of the form $a_1, b_1, a_2, b_2, \ldots, a_k, b_k$, where $a_i \in V_1$ and $b_i \in V_2$ for $i = 1, 2, \ldots, k$. Since the Hamiltonian circuit visits each vertex exactly once, except for v_1, where it begins and ends, the number of vertices in the graph equals $2k$, an even number.

Hence, a bipartite graph with an odd number of vertices cannot have a Hamiltonian circuit.

9. For which values of m and n does the complete bipartite graph $K_{m,n}$ have a Hamiltonian circuit?

Solution. $m = n \geq 2$.

10. In a complete graph with n vertices, show that there are $\dfrac{n-1}{2}$ edge-disjoint Hamiltonian circuits, if n is an odd number ≥ 3.

Solution.
A complete graph G of n vertices has $n(n-1)/2$ edges, and a Hamiltonian circuit in G consists of n edges.

Therefore, the number of edge-disjoint Hamiltonian circuits in G cannot exceed $(n-1)/2$. That is, there are $(n-1)/2$ edge-disjoint Hamiltonian circuits, when n is odd.

The subgraph (of a complete graph of n vertices) in the figure below is a Hamiltonian circuit. Keeping the vertices fixed on a circle, rotate the polygonal pattern clockwise by $360/(n-1)$, $2 \cdot 360/(n-1)$,

$3 \cdot 360/(n-1), \ldots, (n-3)/2 \cdot 360/(n-1)$ degrees. Observe that each rotation produces a Hamiltonian circuit that has no edge in common with any of the previous ones.

Thus, we have $(n-3)/2$ new Hamiltonian circuits, all edge-disjoint from the one as shown below and also edge-disjoint among themselves.

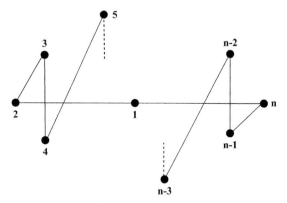

Required graph for the problem

11. Explain Konigsberg bridge problem. Represent the problem by means of a graph. Does the problem have a solution?

Solution.

There are two islands A and B formed by a river. They are connected to each other and to the river banks C and D by means of seven bridges. The problem is to start from any one of the four land areas A, B, C, D, walk across each bridge exactly once, and return to the starting point.

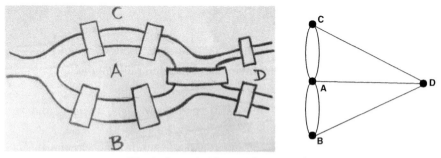

Konigsberg bridge and its graph

The situation is represented by a graph, with vertices representing the land areas and the edges and the bridges.

This problem is the same as that of drawing the graph without lifting the pen from the paper and without retracing any line.

In other words, the problem is to find whether there is an Eulerian circuit in the graph. But a connected graph has an Eulerian circuit if and only if each of its vertices is of even degree.

In the present case, all the vertices are of odd degree. Hence, Konigsberg bridge problem has no solution.

3.6.3 Problems for Practice

1. Can someone cross all the bridges shown in this map exactly once and return to the starting point?

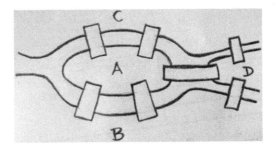

2. In Kaliningrad (the Russian name for Konigsberg), there are two additional bridges, besides the seven that were present in the 18th century. These new bridges connect regions B and C and regions B and D, respectively. Can someone cross all nine bridges in Kaliningrad exactly once and return to the starting point?

3. Show that a directed multigraph having no isolated vertices has an Eulerian circuit if and only if the graph is weakly connected and the in-degree and out-degree of each vertex are equal.

4. For which values of n do the following graphs have an Eulerian circuit?
    (a) K_n (b) C_n (c) W_n (d) Q_n

3.6.4 Additional Problems for Practice

1. Can you draw a graph of five vertices with degree sequence 1, 2, 3, 4, 5?

2. Show that there does not exist a graph with five vertices with degrees 1, 3, 4, 2, 3, respectively.

3. How many edges are there in a graph with ten vertices each of degree 5?

4. Let G be a graph with ten vertices. If four vertices have degree 4 and six vertices have degree 5, then find the number of edges of G.

5. Draw the graph represented by the given adjacency matrix:
$$\begin{pmatrix} 0 & 1 & 0 & 1 \\ 1 & 0 & 1 & 0 \\ 0 & 1 & 0 & 1 \\ 1 & 0 & 1 & 0 \end{pmatrix}.$$

6. How do you find the number of different paths of length r from i to j in a graph G with adjacency matrix A?

7. Is the directed graph given below strongly connected? Why or why not?

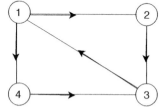

8. Define a bipartite graph.

9. Draw the complete graph K_5.

10. Define isomorphism of directed graphs.

11. What do strongly connected components of a telephone call graph represent?

12. Define Hamiltonian path.

13. Give an example for a graph which is

    (i) Eulerian and Hamiltonian
    (ii) neither Eulerian nor Hamiltonian.

14. Describe a discrete structure based on a graph that can be used to model airline routes and their flight times.

15. Show that a simple graph G with n vertices is connected if it has more than $(n-1)(n-2)/2$ edges.

16. Show the isomorphism of simple graphs is an equivalence relation.

17. Derive an algorithm for constructing Eulerian path in directed graphs.

18. Are simple graphs with the following adjacency matrices isomorphic?

$$\begin{pmatrix} 0 & 1 & 0 & 0 & 0 & 1 \\ 1 & 0 & 1 & 0 & 1 & 0 \\ 0 & 1 & 0 & 1 & 0 & 1 \\ 0 & 0 & 1 & 0 & 1 & 0 \\ 0 & 1 & 0 & 1 & 0 & 1 \\ 1 & 0 & 1 & 0 & 1 & 0 \end{pmatrix}, \begin{pmatrix} 0 & 1 & 0 & 0 & 0 & 1 \\ 1 & 0 & 1 & 0 & 0 & 1 \\ 0 & 1 & 0 & 1 & 1 & 0 \\ 0 & 0 & 1 & 0 & 1 & 0 \\ 0 & 0 & 1 & 1 & 0 & 1 \\ 1 & 1 & 0 & 0 & 1 & 0 \end{pmatrix}$$

19. Examine whether the two graphs G and G' associated with the following adjacency matrices are isomorphic.

$$\begin{pmatrix} 0 & 1 & 0 & 1 & 0 & 0 \\ 1 & 0 & 1 & 0 & 0 & 1 \\ 0 & 1 & 0 & 1 & 0 & 0 \\ 1 & 0 & 1 & 0 & 1 & 0 \\ 0 & 0 & 0 & 1 & 0 & 1 \\ 0 & 1 & 0 & 0 & 1 & 0 \end{pmatrix}, \begin{pmatrix} 0 & 1 & 0 & 0 & 1 & 0 \\ 1 & 0 & 1 & 0 & 0 & 0 \\ 0 & 1 & 0 & 1 & 0 & 1 \\ 0 & 0 & 1 & 0 & 1 & 0 \\ 1 & 0 & 0 & 1 & 0 & 1 \\ 0 & 0 & 1 & 0 & 1 & 0 \end{pmatrix}$$

20. Discuss the various graph invariants preserved by isomorphic graphs.

21. If G is a self-complementary graph, then prove that G has $n \equiv 0$ or $1 \ (mod \ 4)$ vertices.

22. If G is a connected simple graph with n vertices with $n \geq 3$, such that the degree of every vertex in G is at least $\dfrac{n}{2}$, then prove that G has Hamiltonian cycle.

23. In a round robin tournament, the team 1 beats team 2, team 3, and team 4; team 2 beats team 3 and team 4; and team 3 beats team 4. Model this outcome with a directed graph.

24. Show that the number of vertices of odd degree in an undirected graph is even.

25. If a graph, either connected or disconnected, has exactly two vertices of odd degree, prove that there is a path joining these two vertices.

26. Find an Eulerian path or Eulerian circuit if it exists in each of the following two graphs.

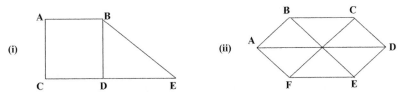

27. Determine whether the following graphs G and H are isomorphic. Give reason.

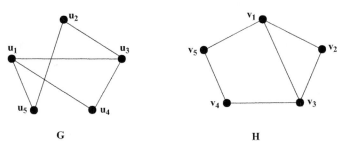

28. Which of the following simple graphs have a Hamiltonian circuit, or if not, a Hamiltonian path?

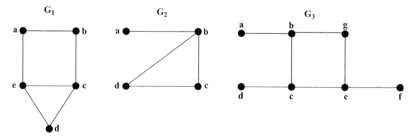

29. Prove that a simple graph is bipartite if and only if it is possible to assign one of two different colours to each vertex of the graph so that no two adjacent vertices are assigned the same colour.

30. Show that every connected graph with n vertices has at least $n-1$ edges.

31. Show that there does not exist a graph with five vertices with degrees 1, 3, 4, 2, 3, respectively.

32. Can you draw a graph of five vertices with degree sequence 1, 2, 3, 4, 5?

33. Draw a graph that is Eulerian but not Hamiltonian.

33. Let G be a graph with ten vertices. If four vertices have degree 4 and six vertices have degree 5, then find the number of edges of G.

34. How many edges are there in a graph with ten vertices each of degree 5?

35. Give an example for a graph which is

    (i) Eulerian and Hamiltonian
    (ii) neither Eulerian nor Hamiltonian.

36. How do you find the number of different paths of length r from i to j in a graph G with adjacency matrix A?

37. Establish the isomorphism of the following pairs of graphs.

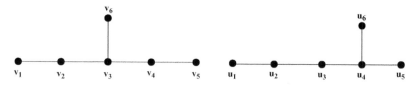

4

Algebraic Structures

4.1 Introduction

An algebraic system can be described as a set of objects together with some operations. These operations will impose a certain structure on the set. In this chapter, we study the axiomatic set theory, semigroups, groups, and monoids which are the basic tools of discrete mathematics.

4.2 Algebraic Systems

Definition 4.2.1 Binary Operation: *Let A be any set. A mapping $f : A \times A \longrightarrow A$ is called a binary operation.*

Definition 4.2.2 Algebraic System or Algebra: *A set together with a number of (binary) operations on the set is called an algebraic system or an algebra.*

Properties of Binary Operations:

Let G be a set.

(i) **Closure Property:** A binary operation $\star : G \times G \longrightarrow G$ is said to be closed, if for all $a, b \in G$, an element $a \star b = x \in G$.

(ii) **Associative Property:** $a \star (b \star c) = (a \star b) \star c$, for all $a, b, c \in G$.

(iii) **Existence of Identity:** There exists an element $e \in G$ such that $e \star a = a \star e = a$, for all $a \in G$. The element e is called the identity element.

(iv) **Existence of Inverse:** For $a \in G$, there exists an element $b \in G$ such that $a \star b = b \star a = e$. The element b is called the inverse of a, and it is denoted by $b = a^{-1}$.

(v) **Commutativity:** For all $a, b \in G$, if $a \star b = b \star a$, then $\star$ is commutative.

(vi) **Distributive Properties:** Let $\bullet$ be any other binary operation on G. Then,

$$a \star (b \bullet c) = (a \star b) \bullet (a \star c) \quad \text{(Left distributive law)}$$
$$(b \bullet c) \star a = (b \star a) \bullet (c \star a) \quad \text{(Right distributive law)}$$

for all $a, b, c \in G$.

(vii) **Cancellation Property:** For $a, b, c \in G$ and $a \neq 0$,

$$a \star b = a \star c \Longrightarrow b = c.$$

Definition 4.2.3 Algebraic Structure: *The operations on a set G define a structure on the elements of G. Then, the algebraic system G is called an algebraic structure.*

Example 4.2.4 *Let $\mathbb{R}$ be the set of real numbers. Consider the algebraic system $(\mathbb{R}, +, \times)$ where $+$ and $\times$ are the operations of addition and multiplication on $\mathbb{R}$.*

4.2.1   Semigroups and Monoids

Definition 4.2.5 Semigroup: *A non-empty set S together with the binary operation $\star : S \times S \longrightarrow S$ is said to be a semigroup if $\star$ satisfies the following conditions, namely, the closure property and the associative property. We denote the semigroup as $(S, \star)$.*

Example 4.2.6 *Let $\mathbb{N} = \{1, 2, 3, \dots\}$ be the set of natural numbers. Then, $(\mathbb{N}, +)$ and $(\mathbb{N}, \bullet)$ are semigroups under the binary operations of addition and multiplication, respectively.*

Definition 4.2.7 Monoid: *A semigroup $(M, \star)$ with an identity element e, with respect to the operation $\star$ is called a monoid. In other words, a non-empty set M together with the binary operation $\star : M \times M \longrightarrow M$ is said to be a monoid if $\star$ satisfies the closure property, associative property, and the identity property.*

Example 4.2.8 *Let $\mathbb{Z}^+ = \{0, 1, 2, 3, \dots\}$ be the set of all non-negative integers. Then, $(\mathbb{Z}^+, +)$ is a semigroup as well as a monoid.*

Definition 4.2.9 *A semigroup (or monoid) $(S, \star)$ is said to be commutative or abelian if $a \star b = b \star a$, for all $a, b \in S$.*

Example 4.2.10 *The set of integers, the set of reals, the set of complex numbers are abelian semigroups (abelian monoids) under the usual operations of addition and multiplication.*

Definition 4.2.11 Idempotent element: *Let $(G, \star)$ be a group. An element $a \in G$ is said to be idempotent if $a \star a = a$.*

Definition 4.2.12 Congruence Relation: *Let $(X, \star)$ be an algebraic system and E an equivalence relation on X. The relation E is called a congruence relation on $(X, \star)$ if E satisfies the substitution property with respect to the operation $\star$.*

Remark 4.2.13 Substitution Property: *Let $(X, \star)$ be an algebraic system in which $\star$ is a binary operation on X. Let us assume that E is an equivalence relation on X.*

The equivalence relation E is said to have the substitution property with respect to the operation $\star$ if and only if for any $x_1, x_2 \in X$,

$$(x_1 E x_1') \wedge (x_2 E x_2') = (x_1 \star x_2) E (x_2 \star x_2')$$

where $x_1', x_2' \in X$.

4.2.2 Solved Problems

1. Show that intersection of any two congruence relations on a set A is again a congruence relation on A.

   Solution.
   Let E_1 and E_2 be two congruence relations on $(A, \star)$.
   $$\implies \quad (a_1 E_1 a_1') \wedge (a_2 E_1 a_2') = (a_1 \star a_2) E_1 (a_1' \star a_2')$$
   $$\text{and} \qquad (a_1 E_2 a_1') \wedge (a_2 E_2 a_2') = (a_1 \star a_2) E_2 (a_1' \star a_2').$$
   Let $E = E_1 \cap E_2$.

   To prove: E is a congruence relation on A

   $$
   \begin{aligned}
   (a_1 E a_1') \wedge (a_2 E a_2') &= [a_1 (E_1 \cap E_2) a_1'] \wedge [a_2 (E_1 \cap E_2) a_2'] \\
   &= (a_1 E_1 a_1') \text{ and } (a_1 E_1 a_1') \wedge (a_2 E_2 a_2') \text{ and } (a_2 E_2 a_2') \\
   &= (a_1 E_1 a_1') \wedge (a_2 E_1 a_2') \text{ and } (a_1 E_2 a_1') \wedge (a_2 E_2 a_2') \\
   &= (a_1 \star a_2) E_1 (a_1' \star a_2') \text{ and } (a_1 \star a_2) E_2 (a_1' \star a_2') \\
   &= (a_1 \star a_2)(E_1 \cap E_2)(a_1' \star a_2') \\
   &= (a_1 \star a_2) E (a_1' \star a_2').
   \end{aligned}
   $$

   Hence, E is a congruence relation on A.

2. Show that a semigroup with more than one idempotent cannot be a group. Give an example of a semigroup which is not a group.

   Solution.
   Let $(S, \star)$ be a semigroup.
   Let a, b are two idempotents.
   $\therefore \quad a \star a = a \quad \text{and} \quad b \star b = b$.
   Let us assume that $(S, \star)$ is group. Then, each element has an inverse. Hence,
   $$(a \star a) \star a^{-1} = a \star (a \star a^{-1}) \quad \text{(by associative property)}.$$

Now,

$$(a \star a) \star a^{-1} = a \star a^{-1} = e \quad \text{since} \quad a \star a = a. \tag{4.1}$$

Also,

$$a \star (a \star a^{-1}) = a \star e = a. \tag{4.2}$$

From (4.1) and (4.2), we get $a = e$.

Similarly, we can prove that $b = e$.

But in a group we cannot have two identities, and hence $(S, \star)$ cannot be a group.

This contradiction is due to an assumption that $(S, \star)$ has two idempotents.

Example: Let $S = \{a, b, c\}$ under the operation $\star$. The composition table of $(S, \star)$ is shown in the following table.

Composition Table of $(S, \star)$

$\star$	a	b	c
a	a	c	a
b	c	b	a
c	b	a	c

$(S, \star)$ is a semigroup which is not a group.

3. Give an example of a semigroup which is not a monoid.

Solution.

Let $D = \{\ldots, -4, -2, 0, 2, 4, \ldots\}$

$(D, \cdot)$ is a semigroup but not a monoid since its multiplicative identity is 1, but $1 \notin D$.

4. Give an example of a monoid which is not a group.

Solution.

$(\mathbb{Z}^+, \cdot)$ is a monoid which is not a group, since for all $a \in \mathbb{Z}^+$, $\dfrac{1}{a} \notin \mathbb{Z}^+$.

5. What do you call a homomorphism of a semigroup into itself?

Solution.

A homomorphism of a semigroup into itself is called a semigroup endomorphism.

6. If $(Z, +)$ and $(E, +)$ are two semigroups, where Z is the set of all integers and E is the set of all even integers, show that the two semigroups $(Z, +)$ and $(E, +)$ are isomorphic.

Solution.
First, we define a function $g : Z \longrightarrow E$ by $g(a) = 2a$, for all $a \in Z$.

To prove g is one-to-one:
Suppose $g(a_1) = g(a_2)$, where $a_1, a_2 \in Z$.
  Then, $2a_1 = 2a_2 \Longrightarrow a_1 = a_2$.
Therefore, g is one-to-one.

To prove g is onto:
Suppose b is an even integer.
  Let $a = \dfrac{b}{2}$. Then, $a \in Z$ and
$$g(a) = g\left(\frac{b}{2}\right) = 2 \cdot \frac{b}{2} = b.$$
That is, every element $b \in E$ has a preimage in Z.
Therefore, g is onto.

To prove g is homomorphism:
Let $a, b \in Z$.
$$g(a + b) = 2(a + b)$$
$$= 2a + 2b$$
$$= g(a) + g(b).$$

Hence, $(Z, +)$ and $(E, +)$ are isomorphic semigroups.

7. If $\star$ is a binary operation on the set of $\mathbb{R}$ of real numbers defined by $a \star b = a + b + 2ab$,

   (i) show that $(\mathbb{R}, \star)$ is a semigroup.
   (ii) find the identity element if it exists.
   (iii) which elements has inverse and what are they?

Solution.

   (i)
$$(a \star b) \star c = (a + b + 2ab) + c + 2(a + b + 2ab)c$$
$$= a + b + c + 2(ab + bc + ca) + 4abc$$

   and
$$a \star (b \star c) = a + (b + c + 2bc) + 2a(b + c + 2bc)$$
$$= a + b + c + 2(ab + bc + ca) + 4abc.$$

   Hence, $(a \star b) \star c = a \star (b \star c)$.
     Therefore, $\star$ is associative.

   (ii) If the identity element exists, let it be e. Then for any $a \in \mathbb{R}$,
$$a \star e = a$$
   or $\qquad a + e + 2ae = a$
   or $\qquad e(1 + 2a) = 0.$

   Therefore, $e = 0$, since $1 + 2a \neq 0$, for any $a \in \mathbb{R}$.

(iii) Let a^{-1} be the inverse of an element $a \in \mathbb{R}$. Then, $a \star a^{-1} = e$.
That is, $a + a^{-1} + 2a \cdot a^{-1} = 0$
$a^{-1} \cdot (1 + 2a) = -a$.
Therefore, $a^{-1} = -\dfrac{a}{1 + 2a}$.

Hence, if $a \neq \dfrac{1}{2}$, then $a^{-1} = -\dfrac{a}{1 + 2a}$.

8. Show that a semigroup with more than one idempotent cannot be a group. Give an example of a semigroup which is not a group.

Solution.

Let $(S, \star)$ be a semigroup.

Let $a, b \in S$ be two idempotents. Then,
$a \star a = a$ and $b \star b = b$.

Let us assume that $(S, \star)$ is a group. Then, each element has its inverse. Now, by associative property, we have

$$(a \star a) \star a^{-1} = a \star (a \star a^{-1}).$$
$$(a \star a) \star a^{-1} = a \star a^{-1} = e. \tag{4.3}$$
$$a \star (a \star a^{-1}) = a \star e = a. \tag{4.4}$$

From (4.3) and (4.4), we get $a = e$.

Similarly, we can prove that $b = e$.

But in a group, we cannot have two identities and hence $(S, \star)$ cannot be a group. This contradiction is due to the assumption that $(S, \star)$ has two idempotents.

Example: Let $S = \{a, b, c\}$ under the operation $\star$. The composition table of $(S, \star)$ is given inthe following table.

Composition Table of $(S, \star)$

$\star$	a	b	c
a	a	c	a
b	c	b	a
c	b	a	c

$(S, \star)$ is a semigroup but not a group.

9. Let $(N, +)$ be the semigroup of natural numbers and $(S, \star)$ be a semigroup where $S = \{e, 0, 1\}$ with the operation $\star$ given in the following table.

Composition Table of $(S, \star)$

$\star$	e	0	1
e	e	0	1
0	0	0	0
1	1	0	1

A mapping $g : N \longrightarrow S$ is defined by $g(0) = 1$ and $g(j) = 0$ for $j \neq 0$. Is g is a semigroup homomorphism?

Solution.

Though both $(N, +)$ and $(S, \star)$ are monoids with identities 0 and e, respectively, g is not a monoid homomorphism because $g(0) \neq e$.

$\therefore$ g is a semigroup homomorphism.

10. If $\star$ is the operation defined on $S = \mathbb{Q} \times \mathbb{Q}$, the set of ordered pair of rational numbers and given by $(a, b) \star (x, y) = (ax, ay + b)$, show that $(S, \star)$ is a semigroup. Is it commutative? Also, find the identity element of S.

Solution.

Given

$$(a, b) \star (x, y) = (ax, ay + b). \tag{4.5}$$

To prove: $(S, \star)$ is a semigroup, that is, to prove $\star$ is associative.

$$[(a, b) \star (x, y)] \star (c, d) = (ax + ay + b) \star (c, d) \quad \text{[using (4.5)]}$$
$$= (acx, adx + ay + b) \quad \text{[using (4.5)]}. \tag{4.6}$$

$$(a, b) \star [(x, y) \star (c, d)] = (a, b) \star (cx, dx + y)$$
$$= (acx, adx + ay + b). \tag{4.7}$$

From (4.6) and (4.7), $\star$ is associative on S.

To prove $(S, \star)$ is not commutative:

$$(x, y) \star (a, b) = (ax, bx + y). \tag{4.8}$$

$$(a, b) \star (x, y) = (ax, ay + b). \tag{4.9}$$

From (4.8) and (4.9), $(a, b) \star (x, y) \neq (x, y) \star (a, b)$.

Hence, $(S, \star)$ is not commutative.

To find the identity element of $(S, \star)$:

Let (e_1, e_2) be the identity element of $(S, \star)$. Then, for all $(a, b) \in S$, we have $(a, b) \star (e_1, e_2) = (a, b)$

$\implies$ $(ae_1, ae_2 + b) = (a, b)$.

$\implies$ $ae_1 = a$ and $ae_2 + b = b$.

$\implies$ $e_1 = 1$ and $ae_2 = 0$ or $e_2 = 0$.

Therefore, $(1, 0)$ is the identity element of $(S, \star)$.

11. Is it true that a semigroup homomorphism preserves identity? Justify your answer. Or prove by an example that semigroup homomorphism need not preserve an identity.

Solution.

To prove that semigroup homomorphism need not preserve an identity:

Let $W = \{0, 1, 2 \ldots\}$. Then, $(W, +)$ is a semigroup homomorphism with identity element 0. Let $S = \{e, 0, 1\}$ and $\star$ be the operation on S given by the table below.

Composition Table of $(S, \star)$

$\star$	e	0	1
e	e	0	1
0	0	0	0
1	1	0	1

Then, $(S, \star)$ is a semigroup with identity e.

Now, define a mapping $g : W \longrightarrow S$ by $g(0) = 1$ and $g(i) = 0$ for $i \neq 0$.

We can see that $g(a + b) = g(a) \star g(b)$, for all $a, b \in W$. Thus, g is a semigroup homomorphism. But $g(0) = 1 \neq e$. Thus, g does not preserve the identity.

12. Find all semigroups of $Z_6, \times_6$ where $Z_6 = \{[0], [1], [2], [3],][4], [5]\}$.

    Solution.

    The composition table is given below.

Composition Table of $(Z_6, \times_6)$

$\times_6$	$[0]$	$[1]$	$[2]$	$[3]$	$[4]$	$[5]$
$[0]$	$[0]$	$[0]$	$[0]$	$[0]$	$[0]$	$[0]$
$[1]$	$[0]$	$[1]$	$[2]$	$[3]$	$[4]$	$[5]$
$[2]$	$[0]$	$[2]$	$[4]$	$[0]$	$[2]$	$[4]$
$[3]$	$[0]$	$[3]$	$[0]$	$[3]$	$[0]$	$[3]$
$[4]$	$[0]$	$[4]$	$[2]$	$[0]$	$[4]$	$[2]$
$[5]$	$[0]$	$[5]$	$[4]$	$[3]$	$[2]$	$[1]$

The semigroups are

$\{[0]\}$,	$\{[0], [1]\}$,	$\{[1]\}$,
$\{[1], [2], [4]\}$,	$\{[0], [1], [2], [4]\}$,	$\{[2], [4]\}$,
$\{[0], [3], [4]\}$,	$\{[1], [5]\}$,	$\{[0], [1], [5]\}$,
$\{[0], [4]\}$,	$\{[0], [1], [4]\}$,	$\{[0], [2], [3]\}$,
$\{[0], [1], [2], [3]\}$.		

13. Prove that $(Z_5, \times_5)$ is a commutative monoid, where $\times_5$ is the multiplication modulo 5.

    Solution.

    $$Z_5 = \{[0], [1], [2], [3], [4]\}.$$

    The composition table is given below.

Composition Table of $(Z_5, \times_5)$

$\times_5$	$[0]$	$[1]$	$[2]$	$[3]$	$[4]$
$[0]$	$[0]$	$[0]$	$[0]$	$[0]$	$[0]$
$[1]$	$[0]$	$[1]$	$[2]$	$[3]$	$[4]$
$[2]$	$[0]$	$[2]$	$[4]$	$[1]$	$[3]$
$[3]$	$[0]$	$[3]$	$[1]$	$[4]$	$[2]$
$[4]$	$[0]$	$[4]$	$[3]$	$[2]$	$[1]$

(i) **Closure Property:**
From the table above, it is clear that Z_5 is closed under $\times_5$.

(ii) **Associative Property:**
It is also clear that the associative property holds from the table above. That is, $[a] \times_5 ([b] \times_5 [c]) = ([a] \times_5 [b]) \times_5 [c]$, for all $[a], [b], [c] \in Z_5$.

(iii) **Existence of Identity:**
$[1]$ is the identity element since $[a] \times_5 [1] = [a]$, for all $[a] \in Z_5$.

(iv) **Commutative Property:**
Clearly, from the table above, $[a] \times_5 [b] = [b] \times_5 [a]$, for all $[a], [b] \in Z_5$.

Hence, $(Z_5, \times_5)$ is a commutative monoid.

14. Let $(M, \star, e_M)$ be a monoid and $a \in M$. If a is invertible, then show that its inverse is unique.

    Solution.
    Let b and c be inverse elements of $a \in M$ such that
    $$a \star b = b \star a = e \text{ and } a \star c = c \star a = e.$$
    Now, $b = b \star e = b \star (a \star c) = (b \star a) \star c = e \star c = c$.
    Therefore, its inverse is unique.

15. Show that the set $\mathbb{N}$ of natural numbers is a semigroup under the operation $x \star y = max\{x, y\}$. Is it a monoid?

    Solution.
    Let $\mathbb{N} = \{1, 2, 3, \dots\}$. Define the operation $x \star y = max\{x, y\}$ for $x, y \in \mathbb{N}$.

    Clearly, $(\mathbb{N}, \star)$ is closed because $x \star y = max\{x, y\} \in \mathbb{N}$ and $\star$ is associative since

    $$(x \star y) \star z = max\{x \star y, z\}$$
    $$= max\{max\{x, y\}, z\}$$
    $$= max\{x, y, z\}$$
    $$= max\{x, max\{y, z\}\} = max\{x, y \star z\} = x \star (y \star z).$$

    Therefore, $(\mathbb{N}, \star)$ is a semigroup.

16. Prove that monoid homomorphism preserves invertibility and monoid epimorphism preserves a zero element (if it exists).

Solution.

Let $(M, \star, e_M)$ and (T, Δ, e_T) be any two monoids, and let $g : M \longrightarrow T$ be a monoid homomorphism. If $a \in M$ is invertible, let a^{-1} be the inverse of a in M. We will now show that $g(a^{-1})$ will be an inverse of $g(a)$ in T.

$$a \star a^{-1} = a^{-1} \star a = e_M \quad \text{(by definition of inverse)}$$

So, $g(a \star a^{-1}) = g(a^{-1} \star a) = g(e_M)$.

Hence, $g(a)\Delta g(a^{-1}) = g(a^{-1})\Delta g(a) = g(e_M)$. (since g is a homomorphism)

But $g(e_M) = e_T$. (since g is a monoid homomorphism)

Therefore, $g(a)\Delta g(a^{-1}) = g(a^{-1})\Delta g(a) = e_T$.

This means $g(a^{-1})$ is an inverse of $g(a)$. That is, $g(a)$ is invertible. Thus, the property of invertibility is preserved under monoid homomorphism.

Assume g is a monoid epimorphism. Let $t = g(b) \in T$. Then

$$t\Delta g(z) = g(b)\Delta g(z) = g(b \star z) = g(z)$$

and $g(z)\Delta t = g(z)\Delta g(b) = g(z \star b) = g(z)$.

Therefore, $g(z)$ is the zero element of T.

17. The operation $\star$ is defined by $a \star b = a + b - ab$, on the set $\mathbb{Q}$ of all rational numbers. Show that under this operation, $\mathbb{Q}$ is a commutative monoid.

Solution.

(i) **Closure Property:**

Since $a + b - ab$ is a rational number for all rational numbers a, b, the given operation $\star$ is a binary operation on $\mathbb{Q}$.

(ii) **Associative Property:**

For all $a, b, c \in \mathbb{Q}$,

$$
\begin{aligned}
(a \star b) \star c &= (a + b - ab) \star c \\
&= (a + b - ab) + c - (a + b - ab)c \\
&= a + b - ab + c - ac - bc + abc \\
&= a + (b + c - bc) - a(b + c - bc) \\
&= a \star (b + c - bc) \\
&= a \star (b \star c).
\end{aligned}
$$

Hence, $\star$ is associative.

(iii) **Existence of Identity:**

For any $a \in \mathbb{Q}$,

$$a \star 0 = a + 0 - a \cdot 0 = a$$

and $0 \star a = 0 + a - 0 \cdot a = a$.

Hence, 0 is the identity element in $\mathbb{Q}$ under the operation $\star$.

(iv) **Commutative Property:**

From the definition of the operation $\star$, it is clear that $\star$ is commutative.

Hence, under the operation $\star$, $\mathbb{Q}$ is a commutative monoid with 0 as the identity element.

18. Let $V = \{a, b\}$ and A be the set of all sequences on V including $\wedge$ beginning with a. Show that $(A, \circ, \wedge)$ is a monoid.

Solution.

Let $V = \{a, b\}$ and A be the set of all sequences on V including $\wedge$ beginning with a. Then, $A = \{\wedge, a, ab, aa, ab, aba, abb, \dots\}$. Let $\circ$ be a concatenation operation on the sequences in A. Clearly for any two elements $\alpha, \beta \in A$, $\alpha \circ \beta = \alpha\beta$ also belongs to A, and hence $(A, \circ)$ is closed.

Also $\circ$ is associative because

$$(\alpha \circ \beta) \circ \gamma = \alpha\beta\gamma$$
$$= \alpha \circ (\beta\gamma)$$
$$= (\alpha \circ \beta \circ \gamma).$$

$\wedge$ is the identity since $\wedge \circ \alpha = \alpha \circ \wedge = \alpha$, for all $\alpha \in A$.

Therefore, $(A, \circ, \wedge)$ is a monoid.

19. Let $V = \{a, b\}$. Show that $(V^\star, \circ, \wedge)$ is an infinite monoid.

Solution.

While defining the alphabet and set of strings $V^\star$, we proved that $(V^\star, \circ, \wedge)$ is a monoid, where $\wedge$ is an empty string. So, it is enough to show that $V^\star$ is an infinite set. As a is an element of V, a, aa, aaa, $aaaa, \dots$; b, bb, bbb, $bbbb, \dots$; and ab, abb, $abbb$, $\dots$ are the elements of $V^\star$, and hence $V^\star$ contains infinitely many strings including an empty set.

4.2.3 Groups

Definition 4.2.14 Group: *A non-empty set G together with a binary operation $\star$, that is $(G, \star)$, is called a group if $\star$ satisfies the following conditions:*

(i) **Associative:** *For every $a, b, c \in G$, $a \star (b \star c) = (a \star b) \star c$.*

(ii) **Existence of Identity:** *There exists an element $e \in G$ called the identity element such that $a \star e = e \star a = a$, for all $a \in G$.*

(iii) **Existence of Inverse:** *There exists an element $a^{-1} \in G$ called the inverse of a such that $a \star a^{-1} = a^{-1} \star a = e$, for each $a \in G$.*

Example 4.2.15 *The set of all integers $\mathbb{Z}$ with the addition operation is a group.*

Example 4.2.16 *The set of all non-zero real numbers $\mathbb{R}^\star$ under the multiplication operation is a group.*

Definition 4.2.17 Abelian Group or Commutative Group: *A group $(G, \star)$ is said to be an abelian group or commutative group if $a \star b = b \star a$, for all $a, b \in G$.*

Otherwise, it is a non-abelian group.

The set of all integers $\mathbb{Z}$ with the addition operation is an abelian group.

Properties of Groups

1. The identity of a group is unique.

2. The left and right cancellation laws are true.

    (i) $a \star b = a \star c \Longrightarrow b = c$ (left cancellation law) and
    (ii) $b \star a = c \star a \Longrightarrow b = c$ (right cancellation law).

3. The inverse of any element in a group is unique.

4. If a is an element of a group G, then $\left(a^{-1}\right)^{-1} = a$.

5. For any two elements a, b in a group G, $(a \star b)^{-1} = b^{-1} \star a^{-1}$.

6. In a group, the solution for the equations $a \star x = b$ and $y \star b = a$ exists, and it is unique.

Theorem 4.2.18 *Every row or column in the composition table of a group $(G, \star)$ is a permutation of the elements of G.*

Proof.

Initially, we shall show that no row or column in the composition table can have an element of G more than once.

Let us assume the contrary. Suppose that the row corresponding to an element $a \in G$ has two entries which are both k. That is, assume that $a \star b_1 = a \star b_2 = k$, where $k, b_1, b_2 \in G$ and $b_1 \neq b_2$. Then by cancellation law, we have $b_1 = b_2$ which is a contradiction. A similar result holds for any column.

Next we will show that every element of G appears in each row and column of the table of composition. Consider the row corresponding to the element $a \in G$, and let b be an arbitrary element of G. Since $b = a \star (a^{-1} \star b)$, "$b$" must appear in the row corresponding to the element $a \in G$. The same argument applies to every column of the table.

Thus, we obtain that no two rows or columns are identical. Hence, every row of the composition table is obtained by a permutation of the elements G and that each row is a distinct permutation. The same result applies to the columns of the composition table.

Theorem 4.2.19 *In a group* $(G, \star)$, *an element* $a \in G$ *such that* $a^2 = e$, $a \neq e$ *if and only if* $a = a^{-1}$.

Proof.

Let us assume that $a = a^{-1}$.

Then, $a^2 = a \star a = a \star a^{-1} = e$.

Conversely, assume that $a^2 = e$ with $a \neq e$.

$$\text{That is,} \quad a \star a = e$$
$$a^{-1} \star a \star a = a^{-1} \star e$$
$$e \star a = a^{-1}$$
$$a = a^{-1}.$$

Theorem 4.2.20 *In a group* $(G, \star)$, $(a^{-1})^{-1} = a$, *for all* $a \in G$.

Proof.

Let a^{-1} be the inverse of a.

$a \star a^{-1} = a^{-1} \star a = e$

$\implies$ a is the inverse of a^{-1}.

That is, $(a^{-1})^{-1} = a$.

Definition 4.2.21 Permutation Group or Symmetric Group: *The set* P_n *of all permutations of* n *elements is a permutation group or a symmetric group under the composition of functions.*

That is, $P_n = \{f / f$ *is a one-to-one and onto mapping from* S *to* $S\}$ *is a group under the composition operation of functions, where* S *is any non-empty set.*

Example 4.2.22 *The set* P_3 *of all permutations on* $S = \{1, 2, 3\}$ *is a finite non-abelian group of order six with respect to composition of mappings.*

The composition table for P_3 *is given in the table below where,*

Composition Table of P_3

$\circ$	f_1	f_2	f_3	f_4	f_5	f_6
f_1	f_1	f_2	f_3	f_4	f_5	f_6
f_2	f_2	f_1	f_6	f_5	f_4	f_3
f_3	f_3	f_5	f_1	f_6	f_2	f_4
f_4	f_4	f_6	f_5	f_1	f_3	f_2
f_5	f_5	f_3	f_4	f_2	f_6	f_1
f_6	f_6	f_4	f_2	f_3	f_1	f_5

$$f_1 = \begin{pmatrix} 1 & 2 & 3 \\ 1 & 2 & 3 \end{pmatrix}, f_2 = \begin{pmatrix} 1 & 2 & 3 \\ 3 & 2 & 1 \end{pmatrix}, f_3 = \begin{pmatrix} 1 & 2 & 3 \\ 2 & 3 & 1 \end{pmatrix}$$

$$f_4 = \begin{pmatrix} 1 & 2 & 3 \\ 3 & 1 & 2 \end{pmatrix}, f_5 = \begin{pmatrix} 1 & 2 & 3 \\ 2 & 1 & 3 \end{pmatrix}, f_6 = \begin{pmatrix} 1 & 2 & 3 \\ 1 & 3 & 2 \end{pmatrix}.$$

4.2.4   Solved Problems

1. Show that $G = \left\{ \begin{pmatrix} a & 0 \\ 0 & 0 \end{pmatrix} : a \neq 0 \in \mathbb{R} \right\}$ is an abelian group under matrix multiplication.

   Solution.

   (i) **Closure Property:**

   Let $A = \begin{pmatrix} a & 0 \\ 0 & 0 \end{pmatrix}, B = \begin{pmatrix} b & 0 \\ 0 & 0 \end{pmatrix} \in G.$

   Then, $AB = \begin{pmatrix} ab & 0 \\ 0 & 0 \end{pmatrix} \in G$ since $ab \in \mathbb{R}$, for all $a, b \in \mathbb{R}$.

   (ii) **Commutative Property:**

   $AB = BA$ is true for all $A, B \in G$, since

   $AB = BA = \begin{pmatrix} ab & 0 \\ 0 & 0 \end{pmatrix}$ $[\because ab = ba$ is true in $\mathbb{R}]$.

   (iii) **Associative Property:**

   Matrix multiplication is associative always. That is, $A(BC) = (AB)C$, for all $A, B, C \in G$.

   (iv) **Existence of Identity:**

   $E = \begin{pmatrix} 1 & 0 \\ 0 & 0 \end{pmatrix} \in G$ is the identity in G, since

   $AE = \begin{pmatrix} a & 0 \\ 0 & 0 \end{pmatrix} \begin{pmatrix} 1 & 0 \\ 0 & 0 \end{pmatrix} = \begin{pmatrix} a & 0 \\ 0 & 0 \end{pmatrix} = A$, for all $A \in G$.

   (v) **Existence of Inverse:**

   If $A = \begin{pmatrix} a & 0 \\ 0 & 0 \end{pmatrix} \in G$, then $A^{-1} = \begin{pmatrix} \frac{1}{a} & 0 \\ 0 & 0 \end{pmatrix} \in G$ is the inverse of

   A, since $AA^{-1} = \begin{pmatrix} 1 & 0 \\ 0 & 0 \end{pmatrix}$ $(\because a \neq 0 \in \mathbb{R} \Longrightarrow \dfrac{1}{a} \neq 0 \in \mathbb{R}).$

   Hence, G is an abelian group under matrix multiplication.

2. Examine whether $G = \left\{ \begin{pmatrix} a & a \\ a & a \end{pmatrix} : a \neq 0 \in R \right\}$ is a commutative group under matrix multiplication, where R is the set of all real numbers.

   Solution.

   (i) **Closure Property:**

   Let $A = \begin{pmatrix} a & a \\ a & a \end{pmatrix}, B = \begin{pmatrix} b & b \\ b & b \end{pmatrix} \in G.$ Then,

   $AB = \begin{pmatrix} 2ab & 2ab \\ 2ab & 2ab \end{pmatrix} \in G$ since $2ab \in R$, for all $a, b \in R$.

   (ii) **Commutative Property:**

   $AB = BA$ is true for all $A, B \in G$, since

   $AB = BA = \begin{pmatrix} 2ab & 2ab \\ 2ab & 2ab \end{pmatrix}$ $[\because 2ab = 2ba$ is true in $R]$.

(iii) **Associative Property:**
Matrix multiplication is associative always. That is,
$A(BC) = (AB)C$, for all $A, B, C \in G$.

(iv) **Existence of Identity:**

$E = \begin{pmatrix} \frac{1}{2} & \frac{1}{2} \\ \frac{1}{2} & \frac{1}{2} \end{pmatrix} \in G$ is the identity in G, since

$$AE = \begin{pmatrix} a & a \\ a & a \end{pmatrix} \begin{pmatrix} \frac{1}{2} & \frac{1}{2} \\ \frac{1}{2} & \frac{1}{2} \end{pmatrix} = \begin{pmatrix} a & a \\ a & a \end{pmatrix} = A, \text{ for all } A \in G.$$

(v) **Existence of Inverse:**

If $A = \begin{pmatrix} a & a \\ a & a \end{pmatrix} \in G$, then $A^{-1} = \begin{pmatrix} \frac{1}{4a} & \frac{1}{4a} \\ \frac{1}{4a} & \frac{1}{4a} \end{pmatrix} \in G$ is the inverse

of A, since $AA^{-1} = \begin{pmatrix} \frac{1}{2} & \frac{1}{2} \\ \frac{1}{2} & \frac{1}{2} \end{pmatrix}$ $\left(\because a \neq 0 \in \mathbb{R} \implies \frac{1}{4a} \neq 0 \in \mathbb{R} \right)$.

Hence, G is a commutative group under matrix multiplication.

3. Prove that $G = \left\{ \begin{pmatrix} 1 & 0 \\ 0 & 1 \end{pmatrix}, \begin{pmatrix} -1 & 0 \\ 0 & 1 \end{pmatrix}, \begin{pmatrix} 1 & 0 \\ 0 & -1 \end{pmatrix}, \begin{pmatrix} -1 & 0 \\ 0 & -1 \end{pmatrix} \right\}$
forms an abelian group under matrix multiplication.

Solution.
Let $A = \begin{pmatrix} 1 & 0 \\ 0 & 1 \end{pmatrix}, B = \begin{pmatrix} -1 & 0 \\ 0 & 1 \end{pmatrix}, C = \begin{pmatrix} 1 & 0 \\ 0 & -1 \end{pmatrix}, D = \begin{pmatrix} -1 & 0 \\ 0 & -1 \end{pmatrix}$.
The composition table is shown in the table below.

Composition Table of $(G, \cdot)$

$\cdot$	A	B	C	D
A	A	B	C	D
B	B	A	D	C
C	C	D	A	B
D	D	C	B	A

(i) **Closure Property:**
Clearly, from the table above, we have $xy \in G$ for all $x, y \in G$.
Hence, closure property is satisfied.

(ii) **Commutative Property:**
We observe from the table above that $xy = yx$, for all $x, y \in G$.
Hence, commutative property holds.

(iii) **Associative Property:**
Matrix multiplication is associative always. That is $x(yz) = (xy)z$,
for all $x, y, z \in G$.

(iv) **Existence of Identity:**

$A = \begin{pmatrix} 1 & 0 \\ 0 & 1 \end{pmatrix}$ is the identity element in G since

$$AA = A, \ AB = BA = B, \ AC = CA = C, \text{ and}$$
$$AD = DA = D.$$

(v) **Existence of Inverse:**

From Table 4.2, all elements in G are self-inverses.

That is, inverse of A is A, inverse of B is B, inverse of C is C, inverse of D is D, since $AA = A$, $BB = A$, $CC = A$, $DD = A$.

Hence, G forms an abelian group under matrix multiplication.

4. Show that $(\mathbb{Q}^+, \star)$ is an abelian group, where $\star$ is defined by $a \star b = \dfrac{ab}{2}, \ \forall \ a, b \in \mathbb{Q}^+.$

Solution.

(i) **Closure Property:**

It is clear that for all $a, b \in \mathbb{Q}^+$, $a \star b \in \mathbb{Q}^+$, since $\dfrac{ab}{2} \in \mathbb{Q}^+$.

Hence, closure property is satisfied.

(ii) **Commutative Property:**

$a \star b = b \star a$ is true for all $a, b \in \mathbb{Q}^+$, since

$$a \star b = b \star a = \frac{ab}{2} \quad \left[\because \frac{ab}{2} = \frac{ba}{2} \text{ is true in } \mathbb{Q}^+ \right].$$

(iii) **Associative Property:**

$$a \star (b \star c) = a \star \left(\frac{bc}{2} \right) = \frac{a \frac{bc}{2}}{2} = \frac{abc}{4}.$$

$$(a \star b) \star c = \left(\frac{ab}{2} \right) \star c = \frac{\frac{ab}{2} c}{2} = \frac{abc}{4}.$$

Therefore, $a \star (b \star c) = (a \star b) \star c$, for all $a, b, c \in \mathbb{Q}^+$.

Hence, associative property is satisfied.

(iv) **Existence of Identity:**

$e = 2 \in \mathbb{Q}^+$ is the identity element, since

$$a \star e = a \star 2 = \frac{a \cdot 2}{2} = a, \text{ for all } a \in \mathbb{Q}^+.$$

(v) **Existence of Inverse:**

$a^{-1} = \dfrac{4}{a} \in \mathbb{Q}^+$ is the inverse of $a \in \mathbb{Q}^+$, since

$$a \star a^{-1} = a \star \frac{4}{a} = \frac{a \cdot \frac{4}{a}}{2} = \frac{4a}{2a} = 2.$$

Hence, $\mathbb{Q}^+$ is an abelian group under the operation $\star$ defined in the problem.

5. Prove that the identity element of a group is unique.

Solution.

Let $(G, \star)$ be a group.

Let e_1 and e_2 be two identity elements in G.

Then,

$$e_1 \star e_2 = e_1 \qquad [\because e_2 \text{ is the identity}]$$
$$e_1 \star e_2 = e_2 \qquad [\because e_1 \text{ is the identity}].$$

Thus, $e_1 = e_2$.

Hence, the identity is unique.

6. Prove that the identity element is the only idempotent element of a group.

Solution.

Let $(G, \star)$ be a group.

Since $e \star e = e$, e is the idempotent element.

Let a be any idempotent element of G.

Then, $a \star a = a$.

Also, $e \star a = a$ $[\because e$ is the identity element$]$.

It follows that $a \star a = e \star a$.

By the right cancellation law, we have $a = e$, and so e is the only idempotent element.

7. Prove that if every element in a group is its own inverse, then the group must be abelian. Or prove that for any group $(G, \star)$, if $a^2 = e$ with $a \neq e$, then G is abelian.

Solution.

Given $a = a^{-1}$ for all $a \in G$.

Let $a, b \in G$. Then, $a = a^{-1}$ and $b = b^{-1}$.

Now,
$$(a \star b) = (a \star b)^{-1}$$
$$= b^{-1} \star a^{-1}$$
$$= b \star a.$$
$$\implies \quad G \text{ is abelian.}$$

8. Prove for any element a in a group G, the inverse is unique.

Solution.

Let a be any element of a group G.

If possible, let a' and a'' be two inverses of a. Then
$$a \star a' = a' \star a = e$$
$$a \star a'' = a'' \star a = e.$$
Now, $a' = a' = a' \star (a \star a'') = (a' \star a) \star a'' = e \star a'' = a''$.

Hence, the inverse is unique.

9. Prove that in a group $(G, \star)$, $(a \star b)^{-1} = b^{-1} \star a^{-1}$.

Solution.

$$(a \star b)(b^{-1} \star a^{-1}) = a \star (b \star b^{-1}) \star a^{-1}$$

$$= a \star e \star a^{-1}$$
$$= a \star a^{-1} = e$$

and

$$(b^{-1} \star a^{-1}) \star (a \star b) = b^{-1} \star a^{-1} \star a \star b$$
$$= b^{-1} \star e \star b$$
$$= b^{-1} \star b = e.$$

Hence, $(a \star b)^{-1} = b^{-1} \star a^{-1}$.

10. If a and b are any two elements of a group $(G, \star)$, then show that G is abelian if and only if $(a \star b)^2 = a^2 \star b^2$.

Solution.
Necessary Part:
Given that $(G, \star)$ is an abelian group.

$$\Longrightarrow \qquad \text{For all} \quad a, b \in G, a \star b = b \star a. \tag{4.10}$$

To prove: $(a \star b)^2 = a^2 \star b^2$.

$$(a \star b)^2 = (a \star b) \star (a \star b)$$
$$= a \star (b \star a) \star b$$
$$= a \star (a \star b) \star b \quad [\text{using (4.10)}]$$
$$= (a \star a) \star (b \star b)$$
$$= a^2 \star b^2.$$

Sufficient Part:

$$\text{Given:} \qquad (a \star b)^2 = a^2 \star b^2. \tag{4.11}$$

To prove: $a \star b = b \star a$.

$$(4.11) \Longrightarrow \qquad (a \star b)^2 = a^2 \star b^2$$
$$\Longrightarrow \qquad (a \star b) \star (a \star b) = (a \star b) \star (b \star b)$$
$$\Longrightarrow \qquad a \star [b \star (a \star b)] = a \star [a \star (b \star b)]$$
$$\Longrightarrow \qquad b \star (a \star b) = a \star (b \star b) \quad (\text{using left cancellation law})$$
$$\Longrightarrow \qquad (b \star a) \star b = (a \star b) \star b \quad (\text{using associative property})$$
$$\Longrightarrow \qquad b \star a = a \star b \quad (\text{using right cancellation law})$$
$$\Longrightarrow \qquad G \text{ is abelian.}$$

11. Show that every group of order four is abelian.

Solution.
Let $(G, \star)$ be a group of order four where $G = \{e, a, b, c\}$. Since G is of even order, there exists at least one element, say a such that $a^{-1} = a$. Then, two cases arise:

(i) $b^{-1} = b$, $c^{-1} = c$

(ii) $b^{-1} = c$, $c^{-1} = b$.

Case (i): $e^{-1} = e$, $a^{-1} = a$, $b^{-1} = b$, $c^{-1} = c$.

That is, every element has its own inverse.

Then, $(G, \star)$ is abelian.

Case (ii): $a^{-1} = a$, $b^{-1} = c$, $c^{-1} = b$.

Therefore, $a^2 = e$, $b \star c = e$, $c \star b = e$.

Since $(G, \star)$ is a group, its elements will appear in a row (column) only once.

Since a, e appear in the second row and b appears in the third column, c will appear as $(2, 3)^{\text{th}}$ element.

$\therefore$ $(2, 4)^{\text{th}}$ element is b

$(3, 3)^{\text{th}}$ element is a

$(3, 2)^{\text{th}}$ element is c

$(4, 2)^{\text{th}}$ element is b

$(4, 4)^{\text{th}}$ element is a.

Composition Table of $(G, \star)$

$\star$	e	a	b	c
e	e	a	b	c
a	a	e	c	b
b	b	c	a	e
c	c	b	e	a

12. Show that the set $S = \{[1], [5], [7], [11]\}$ is a group with respect to multiplication modulo 12.

Solution.

The composition table of S with respect to $\times_{12}$ is given in the table below: Here, $5 \times_{12} 7 = 35$, which on division by 12 gives the

Composition Table of $(S, \times_{12})$

$\times_{12}$	$[1]$	$[5]$	$[7]$	$[11]$
$[1]$	$[1]$	$[5]$	$[7]$	$[11]$
$[5]$	$[5]$	$[1]$	$[11]$	$[7]$
$[7]$	$[7]$	$[11]$	$[1]$	$[5]$
$[11]$	$[11]$	$[7]$	$[5]$	$[1]$

remainder 11, $11 \times_{12} 7 = 77$, which on division by 12 gives the remainder 5, etc.

Hence, S is a group, in which $[1]$ is the identity and each element of S is its own inverse.

13. Show the set of matrices $G = \left\{ \begin{pmatrix} \cos \alpha & -\sin \alpha \\ \sin \alpha & \cos \alpha \end{pmatrix}, a \in \mathbb{R} \right\}$ forms a group under matrix multiplication.

Solution.

(i) **Closure Property:**

Let $A_\alpha = \begin{pmatrix} \cos\alpha & -\sin\alpha \\ \sin\alpha & \cos\alpha \end{pmatrix} \in G$ and $A_\beta = \begin{pmatrix} \cos\beta & -\sin\beta \\ \sin\beta & \cos\beta \end{pmatrix} \in G$.

Then

$$A_\alpha A_\beta = \begin{pmatrix} \cos\alpha & -\sin\alpha \\ \sin\alpha & \cos\alpha \end{pmatrix} \begin{pmatrix} \cos\beta & -\sin\beta \\ \sin\beta & \cos\beta \end{pmatrix}$$

$$= \begin{pmatrix} \cos\alpha\cos\beta - \sin\alpha\sin\beta & -(\cos\alpha\sin\beta + \sin\alpha\cos\beta) \\ \sin\alpha\cos\beta + \cos\alpha\sin\beta & \cos\alpha\cos\beta - \sin\alpha\sin\beta \end{pmatrix}$$

$$= \begin{pmatrix} \cos(\alpha+\beta) & -\sin(\alpha+\beta) \\ \sin(\alpha+\beta) & \cos(\alpha+\beta) \end{pmatrix} = A_{\alpha+\beta} \in G. \tag{4.12}$$

(ii) **Associative Property:**

We know that matrix multiplication is associative.

(iii) **Existence of Identity:**

$I_0 = \begin{pmatrix} 1 & 0 \\ 0 & 1 \end{pmatrix}$ is the identity in G, since $A_\alpha I_0 = I_0 A_\alpha = A_\alpha$ for $A_\alpha \in G$.

(iv) **Existence of Inverse:**

$A_{-\alpha}$ is the inverse of A_α for each $A_\alpha \in G$,

since $A_\alpha A_{-\alpha} = A_{\alpha+(-\alpha)} = A_0 = I_0$, using (4.12)

Hence, G forms a group under matrix multiplication.

4.2.5   Subgroups

Definition 4.2.23 Subgroup: *A non-empty subset H of a group G is said to be a subgroup of G, if H itself is a group under the same operation defined on G and with the same identity element.*

Example 4.2.24 *The set of all integers $\mathbb{Z}$ is a subgroup of the set of all real numbers $\mathbb{R}$ under usual addition. That is, $(\mathbb{Z}, +)$ is a subgroup of $(\mathbb{R}, +)$.*

Theorem 4.2.25 *The necessary and sufficient condition is that a non-empty subset H of a group $(G, \star)$ is a subgroup iff for any $a, b \in H$, $a \star b^{-1} \in H$.*

Proof.

Necessary condition:

Assume that H is a subgroup of G.

Since H itself is a group, we have $a, b \in H \implies a \star b \in H$ (using closure property). Also, $b \in H \implies b^{-1} \in H$ (using inverse property).

$\therefore \quad a, b \in H \implies a, b^{-1} \in H \implies a \star b^{-1} \in H$.

Sufficient condition:

Let $a \star b^{-1} \in H$, for all $a, b \in H$ and H is a subset of G.

We have to prove H is a subgroup of G.

(i) **Existence of Identity:**
Let $a \in H \implies a \star a^{-1} \in H \subseteq G$
$$\implies e \in H.$$
$\therefore$ e is the identity element of H.

(ii) **Existence of Inverse:**
Let $e \in H, a \in H$
$$\implies e \star a^{-1} \in H \subseteq G$$
$$\implies a^{-1} \in H.$$
$\therefore$ Every element of H has an inverse in H.

(iii) **Closure Property:**
Let $b \in H \implies b^{-1} \in H.$

$$\therefore \quad \text{For} \quad a, b \in H \implies a, b^{-1} \in H$$
$$\implies a \star \left(b^{-1} \right)^{-1} \in H \subseteq G$$
$$\implies a \star b \in H.$$

$\therefore$ H is closed under the operation $\star$.

(iv) **Associative Property:**
Given that $H \subseteq G$.
$\implies$ The elements of H are also the elements of G. Since $\star$ is associative in G, it must also be associative in H.
Therefore, H itself is a group under the operation $\star$ in G.

4.2.6 Cyclic Groups

Definition 4.2.26 Order of a group: *Let $(G, \star)$ be a group. The number of elements in G is called the order of the group G and is denoted by $O(G)$.*

Note: If $O(G)$ is finite, then G is called a finite group, otherwise it is called an infinite group.

Definition 4.2.27 Cyclic Group: *A group $(G, \star)$ is said to be cyclic if there exists an element $a \in G$ such that any $x \in G$ can be written as either $x = a^n$ or $x = na$, where n is some integer.*
This element a is called the generator of the cyclic group G, that is, the cyclic group generated by a, and we denote it by $G = <a>$.

Example 4.2.28 *The multiplicative group, $G = \{1, -1, i, -i\}$, (i being the complex number) is cyclic.*
We can write $1 = i^4$, $-1 = i^2$, $i = i^3$. That is all the elements of G can be expressed as integral powers of the element i.
Therefore, G is a cyclic group generated by i. Since i is the generator of G, i^{-1} is also a generator of G.
Hence, G is a cyclic group, and its generators are i and i^{-1}.

Theorem 4.2.29 *Every cyclic group is abelian.*

Proof.
Let $(G, \star)$ be a cyclic group generated by an element $a \in G$, that is $G = <a>$.

Then, for any two elements $x, y \in G$, we have $x = a^n, y = a^m$, where m, n are integers. Therefore,

$$x \star y = a^n \star a^m$$
$$= a^{m+n}$$
$$= a^m \star a^n$$
$$= y \star x.$$

Thus, $(G, \star)$ is abelian.

Theorem 4.2.30 *Let $(G, \star)$ be a finite group generated by an element $a \in G$. If G is of order n, that is, $O(G) = n$, then $a^n = e$ so that $G = \{a, a^2, \ldots, a^n = e\}$. Further, n is the least positive integer for which $a^n = e$.*

Proof.
Let us assume that, for some positive integer $m < n$, $a^m = e$.

Since G is cyclic, any element of G can be written as a^k, for some $k \in \mathbb{Z}$. By division algorithm, we have $k = mq + r$, where $q \in \mathbb{Z}$ and $0 \leq r \leq m$. Therefore,

$$a^k = a^{mq+r}$$
$$= a^{mq} \star a^r$$
$$= (a^m)^q \star a^r$$
$$= e^q \star a^r$$
$$= e \star a^r$$
$$= a^r.$$

Hence, every element of G can be expressed as a^r, for some $0 \leq r \leq m$.

Therefore, G has at most m distinct elements. That is, $O(G) = m < n$, which is a contradiction.

Hence, $a^m = e$, for $m < n$ is not possible.

We now proceed to show that the elements $a, a^2, a^3, \ldots, a^n$ are all distinct where $a^n = e$.

If possible, let $a^i = a^j$, for $i < j \leq n$. Therefore,

$$a^i \star a^{-j} = a^j \star a^{-j}$$
$$\implies \quad a^{i-j} = a^{j-j} = e, \quad \text{where } i - j < n,$$

which is again a contradiction.

Hence, $a^i \neq a^j$, for $i, j \leq n$.

Hence, the theorem is proved.

Theorem 4.2.31 *Every subgroup of a cyclic group is cyclic.*

Proof.
Let G be a finite cyclic group of order n with generator a. That is,

$$g = \{e, a, a^2, \ldots, a^{n-1}\}.$$

Let H be a subgroup of G. Then, elements of H are of the form a^k with $1 \leq k < n$. Let t be the smallest positive integer such that $a^t \in H$. We shall prove that $H = <a^t>$. Indeed, let $a^m \in H$. By the division algorithm, there exist unique integers q and r such that $m = tq + r$ where $0 \leq r < t$. It follows that $a^m = (a^t)^q a^r$ or $a^r = a^m (a^t)^{-q}$. But $a^m \in H$ and $a^t \in H$. Then by closure, $a^r \in H$. Since t is the smallest positive integer such that $a^t \in H$, we must have $r = 0$. Hence, $a^m = (a^t)^q$ or $q^m \in <a^t>$. Clearly, $<a^t> \subseteq H$ since $a^t \in H$ and H is a group.

Theorem 4.2.32 *Every group of prime order is cyclic and hence is abelian.*

Proof.
Let G be a group with $O(G) = p$, a prime.
Let $a \neq e \in G$ and $H = <a>$ be the cyclic group of G generated by a.
By Lagrange's theorem, $O(H)|p$. So, $O(H) = 1$ or p.
Since $O(H) \neq 1$ (as $a \neq e$ and $a, e \in H$, $O(H) \geq 2$), we have $O(H) = p$.
So, $G = H = <a>$ is a cyclic group.
Since every cyclic group is abelian, G is abelian.

4.2.7 Homomorphisms

Definition 4.2.33 Homomorphism: *Let* $(G, \star)$ *and* (H, Δ) *be any two groups. A mapping* $f : G \longrightarrow H$ *is said to be a homomorphism if* $f(a \star b) = f(a) \Delta f(b)$, *for* $a, b \in G$.

Example 4.2.34 *Let* $G = (\mathbb{Z}, +)$ *and* $H = (n\mathbb{Z}, +)$ *be two groups (for a fixed integer n). The mapping* $f : G \longrightarrow H$ *defined by* $f(m) = nm$ *for* $m \in \mathbb{Z}$ *is a homomorphism from G into H.*

Definition 4.2.35 Kernel of a Homomorphism: *Let* $f : G \longrightarrow G'$ *be a group homomorphism. The set of elements of G which are mapped into e' (the identity of G') is called the kernel of f and is denoted by $ker(f)$. That is,*

$$ker(f) = \{x \in G / f(x) = e', \text{ where } e' \text{ is the identity of } G'\}.$$

Theorem 4.2.36 *The kernel of a homomorphism f from a group G into a group G' is a subgroup of G.*

Proof.
Let $f : (G, \star) \longrightarrow (G', \star')$ be any homomorphism.

Then, $ker(f) = \{x \in G / f(x) = e', \text{ where } e' \text{ is the identity of } G'\}.$

Since $f(e) = e'$ is true always, at least $e \in ker(f)$. Therefore, $ker(f)$ is a non-empty subset of G.

Let $a, b \in ker(f)$ with $f(a) = e'$ and $f(b) = e'$. Therefore,

$$f(a \star b^{-1}) = f(a) \star' f(b^{-1}) \text{(since } f \text{ is a homomorphism)}$$
$$= f(a) \star' (f(b))^{-1}$$
$$= e' \star' e'$$
$$= e'$$
$$\implies \quad a \star b^{-1} \in ker(f).$$

That is, $a, b \in ker(f) \implies a \star b^{-1} \in ker(f)$. Hence, $ker(f)$ is a subgroup of G.

Definition 4.2.37 Endomorphism: *A homomorphism f of a group into itself is called an endomorphism.*

Definition 4.2.38 Isomorphism: *A mapping f from a group G to a group G' is said to be an isomorphism if f is a one-to-one and onto homomorphism.*

Theorem 4.2.39 Cayley's Representation Theorem: *Every finite group of order n is isomorphic to a permutation group of degree n.*

Proof.
Let G be any finite group of order n. For each $a \in G$, define a function $f_a : G \longrightarrow G$ such that $f_a(x) = ax$, for every $x \in G$.

Clearly, this function f_a is bijective (one-to-one and onto).

Consider $G_1 = \{f_a / a \in G\}$.

This G_1 becomes a group under the composition operation of functions.

That is, $(G_1, \circ)$ is the permutation group of order n.

Now, define a function $\Phi : G \longrightarrow G_1$ such that $\Phi(a) = f_a$, for all $a \in G$.

Claim 1: Φ is a homomorphism

$$\Phi(b) = f_{ab}$$
$$= f_a \circ f_b \quad [\text{since } f_{ab}(x) = abx = a(bx) = f_a(bx) = f_a \circ f_b(x)]$$
$$= \Phi(a) \circ \Phi(b).$$

Claim 2: Φ is bijective
Clearly, Φ is one-to-one, since

$$\Phi(a) = \Phi(b)$$
$$\implies \quad f_a = f_b$$
$$\implies \quad f_a(x) = f_b(x), \quad \text{for every } x \in G$$
$$\implies \quad ax = bx$$
$$\implies \quad a = b.$$

Also, for every $f_a \in G$, we have $a \in G$ such that $\Phi(a) = f_a$.

Therefore, Φ is onto.

Hence, Φ is bijective. Thus, $\Phi : G \longrightarrow G_1$ becomes as an isomorphism.

Hence, every finite group of order n is isomorphic to a permutation group of degree n.

Theorem 4.2.40 *Any cyclic group of order n is isomorphic to the additive group of residue classes of integers modulo n.*

Proof.

Let $G = \{a, a^2, \ldots, a^n = e\}$ be a cyclic group of order n generated by a.

We know that $(Z_n, +_n)$ is the additive group of residue classes modulo n

$\implies \quad Z_n = \{[1], [2], \ldots, [n] = [0]\}$.

Let $f : G \longrightarrow Z_n$ be defined by $f(a^r) = [r]$, for all $a^r \in G$.

For all $[r] \in Z_n$, there exists $a^r \in G$ such that $f(a^r) = [r]$

$\implies \quad f$ is onto.

For $r \neq s$, $[r] \neq [s]$ and hence $f(a^r) \neq f(a^s)$

$\implies \quad f$ is one-to-one.

For all $a^r, a^s \in G$, $f(a^r \cdot a^s) = f(a^{r+s}) = [r + s] = [r] + [s]$

$$= f(a^r) +_n f(a^s)$$

$\implies \quad f$ is a homomorphism.

Hence, $(G, \cdot)$ is isomorphic to $(Z_n, +_n)$.

4.2.8 Cosets and Normal Subgroups

Definition 4.2.41 Left and Right Cosets: *Let $(H, \star)$ be a subgroup of a group $(G, \star)$.*

(i) *For any $a \in G$, the set $a \star H$ defined by*

$$a \star H = \{a \star h / h \in H\}$$

is called the left coset of H in G determined by the element $a \in G$.

(ii) *For any $a \in G$, the set $H \star a$ defined by*

$$H\star = \{h \star a / h \in H\}$$

is called the right coset of H in G determined by the element $a \in G$.

Example 4.2.42 *Consider the multiplicative group $G = \{1, -1, i, -i\}$ and a subgroup $H = \{1, -1\}$. Clearly, iH, $-iH$, $1H$, and $-1H$ are the left cosets.*

Definition 4.2.43 Index of a subgroup in a group: *Let $(H, \star)$ be a subgroup of a group $(G, \star)$. Then, the number of different left (or right) cosets of H in G is called the index of H in G, and it is denoted by $i_G(H)$.*

Some important results:

(i) If G is abelian, then $a \star H = H \star a$, for all $a \in G$.

(ii)   If H is a subgroup of G and $e \in H$, then $e \star H = H \star e = H$.

(iii)   Any two left or right cosets of H in G are either disjoint or identical.

(iv)   The union of all distinct left (or right) cosets of H in G is equal to G.

Theorem 4.2.44 *Let $(H, \star)$ be a subgroup of a group $(G, \star)$. The set of left cosets of H in G forms a partition of G. Also, every element of G belongs to one and only one left coset of H in G.*

Proof.
To prove: Every element of G belongs to one and only one left coset of H in G.

Let H be a subgroup of a group G. Let $a \in G$. Then, $aH = H$ if and only if $a \in H$.

Suppose $a \in G$ and $aH = H$. Then,

$$aH = H \Longrightarrow ae \in H \Longrightarrow a \in H \quad \text{(since } H \text{ is a subgroup and } e \in H\text{)}.$$

Conversely, assume that $a \in H$.
Then $ah \in H$, for all $h \in H$. So,

$$aH \subseteq H. \tag{4.13}$$

Given any $y \in H$, $a^{-1}y \in H$, and $y = a(a^{-1}y) \in H$. So, $y \in aH$, for all $y \in H$. That is,

$$H \subseteq aH. \tag{4.14}$$

From (4.13) and (4.14), we have

$$H = aH.$$

Hence, every element of G belongs to one and only one left coset of H in G.
To prove: The set of left cosets of H in G forms a partition of G.

Let $a, b \in G$ and H be a subgroup of G.
If $aH \cap Ha \neq \phi$, then let $c \in aH \cap Ha$.
Since $c \in aH$, we have $cH = aH$.
    Let H be a subgroup of a group G. Let $a, b \in G$ if $b \in aH$; then $bH = aH$.
Since $c \in bH$, we have $cH = bH$. So

$$aH = cH = bH.$$

Thus, if $aH \cap bH \neq \phi$, then $aH = bH$.
    Therefore, any two distinct left cosets are disjoint. Hence, the set of all (distinct) left cosets of H in G forms a partition of G.

Theorem 4.2.45 Lagrange's Theorem: *The order of each subgroup of a finite group is a divisor of the order of the group.*

Proof.

Let G be a finite group and H be a subgroup of G.

Let $O(G) = n$ and $O(H) = m$. Let us consider all left cosets of H in G.

Each coset has exactly m elements.

$ah_1 = ah_2 \implies h_1 = h_2$, for all $a \in G$.

By result (iii), namely, G is decomposed into say r mutually disjoint subsets, each of order m.

Therefore, $n = rm$. That is, $O(G) = rO(H)$.

Thus, $O(H)$ divides $O(G)$.

Note: The converse of Lagrange's theorem is not true in general. That is, if n is a divisor of a group G, then it does not necessarily follow that G has a subgroup of order n.

Theorem 4.2.46 *If $(G, \star)$ is a finite group of order n, then for any $a \in G$, we must have $a^n = e$, where e is the identity of the group G.*

Proof.

Let $O(G) = n$. Let $a \in G$.

Then, the order of the subgroup $<a>$ is the order of the element a.

If $O(<a>) = m$, then $a^m = e$, and by Lagrange's theorem, we get $m|n$.

Let $n = mk$. Then, $a^n = a^{mk} = (a^m)^k = e^k = e$.

Definition 4.2.47 Normal Subgroup: *A subgroup $(H, \star)$ of a group $(G, \star)$ is said to be a normal subgroup of G if for every $x \in G$ and for every $h \in H$, $xhx^{-1} \in H$ or $xHx^{-1} \subseteq H$.*

Example 4.2.48 *Consider the group $(\mathbb{Z}, +)$. Clearly, $(3\mathbb{Z}, +)$ is a normal subgroup of $(\mathbb{Z}, +)$.*

Definition 4.2.49 Quotient Group or Factor Group: *Let N be a normal subgroup of a group $(G, \star)$ and the set of all right cosets of N in G be denoted by*

$$G/N = \{Na | a \in G\}.$$

Now, define $\otimes$ as binary operation on G/N as

$$Na \otimes Nb = N(a \star b).$$

Then, $(G/N, \otimes)$ will form a group called quotient group or factor group.

Theorem 4.2.50 *The kernel of a homomorphism is a normal subgroup.*

Proof.

Let $f : (G, \star) \longrightarrow (G', \star')$ be any homomorphism.

Then, $ker(f) = \{x \in G / f(x) = e'$ (where e' is the identity element of $G')\}$ is a subgroup of G by Theorem 4.2.36.

Let $x \in ker(f)$ and let $g \in G$.

$\implies \quad f(x) = e'$.

Consider

$$f(g \star x \star g^{-1}) = f(g) \star' f(x \star g^{-1})$$
$$= f(g) \star' [f(x) \star' f(g^{-1})]$$
$$= f(g) \star' [e' \star' f(g^{-1})]$$
$$= f(g) \star' f(g^{-1})$$
$$= f(g \star g^{-1})$$
$$= f(e) = e'.$$

Thus, $f(g \star x \star g^{-1}) = e'$.

Therefore, $g \star x \star g^{-1} \in ker(f)$.

Hence, $ker(f)$ is a normal subgroup.

Theorem 4.2.51 *Let $(H, \star)$ be a subgroup of a group $(G, \star)$. Then, $(H, \star)$ is a normal subgroup if and only if $a \star h \star a^{-1} = H$, for all $a \in G$.*

Proof.

Let H be a normal subgroup of G.

Then by definition, $a \star H = H \star a$, for all $a \in G$. Hence,

$$a \star H \star a^{-1} = a \star (a^{-1} \star H)$$
$$= (a \star a^{-1}) \star H$$
$$= e \star H$$
$$= H.$$

Conversely, let $a^{-1} \star H \star a = H$, for all $a \in G$.

That is, $a \star (a^{-1} \star H \star a) = a \star H$

or $(a \star a^{-1}) \star (H \star a) = a \star H$

or $e \star (H \star a) = a \star H$

or $H \star a = a \star H.$

Thus, H is a normal subgroup.

Theorem 4.2.52 *Let $(G, \star)$ be a group.*
Let $H = \{a | a \in G \ \& \ a \star b = b \star a, \forall \ b \in G\}$. Then, H is a normal subgroup.

Proof.

$H = \{a | a \in G \ \& \ a \star b = b \star a, \forall \ b \in G\}$.

Since $e \star a = a \star e = a, \forall a \in G$, we have $e \in H$.

Therefore, H is non-empty.

Let $x, y \in H$. Then

$$a \star x = x \star a, \forall \ x \in G, \text{ and } a \star y = y \star a, \forall \ y \in G.$$

Claim: H is a normal subgroup. Consider

$$a \star (x \star y) = (a \star x) \star y$$
$$= (x \star a) \star y$$
$$= x \star (a \star y)$$

$$= x \star (y \star a)$$
$$= (x \star y) \star a$$
$$\implies x \star y \in H.$$

Let $a \in H$. Then, $a \star x = x \star a$, $\forall x \in G$. Then

$$a^{-1} \star (a \star x) = a^{-1} \star (x \star a)$$
$$\implies x = a^{-1} \star (x \star a)$$
$$\implies x \star a^{-1} = a^{-1} \star (x \star a) \star a^{-1}$$
$$= (a^{-1} \star x) \star (a \star a^{-1})$$
$$= a^{-1} \star x$$
$$\implies x \star a^{-1} = a^{-1} \star x, \ \forall x \in G$$
$$\implies a^{-1} \in H.$$

Thus, H is a subgroup.

To prove: H is normal.
Let $x \in H$, $g \in G$.
$\quad$ Then, $a \star x = x \star a$, $\forall a \in G$.
$\quad$ Then, $g \star x \star g^{-1} = x \star g \star g^{-1}$
$\quad \implies x \in H.$
$\quad$ Thus, $g \star x \star g^{-1} \in H \implies H$ is normal.

Theorem 4.2.53 *N is a normal subgroup of a group G if and only if $gNg^{-1} = N$, for every $g \in G$ (or $gN = Ng$). Show that the number of right and left cosets are equal in normal subgroups and every left coset is a right coset.*

Proof.
Let N be a normal subgroup of G.
$\quad$ Let $x \in gNg^{-1} \implies x = gng^{-1}$, for some $n \in N$.
$\qquad$ Therefore, $x = gng^{-1} \in N$ $\quad$ ($\because$ N is a normal subgroup).
$\qquad$ Hence, $gNg^{-1} \subseteq N$.
Now, $\quad g^{-1}Ng = g^{-1}N(g^{-1})^{-1} \subseteq N$, since $g^{-1} \in G$, and $g^{-1}ng \in N$.
$\qquad$ Therefore, $N = g(g^{-1}Ng)g^{-1} \in gNg^{-1}$
$\qquad$ Therefore, $N \subseteq gNg^{-1}$.
$\qquad$ Hence, $N = gNg^{-1}$.
Conversely, let $Ng^{-1} = N$, for every $g \in G$.
$\quad$ That is, gNg^{-1} is the set of all gng^{-1}, for $n \in N$.
$\quad$ Clearly, $gNg^{-1} \subseteq N$.
$\qquad$ Therefore, N is a normal subgroup.
$\quad$ We get if N is a normal subgroup, then $gNg^{-1} = N$ or $gN = Ng$, that is, the left and right cosets are equal.
$\quad$ Therefore, the right and left cosets are equal in number in normal subgroups, and every left coset is a right coset.

Fundamental Theorem of Group Homomorphism

Let H be a normal subgroup of a group G. Let G/H be the set of all left cosets of H in G. That is, $G/H = \{aH/a \in G\}$. Let us define an operation "." as follows.

For any $a, b \in G$, $(ab)H = (aH)(bH)$.

This is a binary operation under which G/H becomes a group. It is called the quotient group or factor group.

Let $f : G \longrightarrow G/H$ be defined as $f(a) = aH$, for any $a \in G$.

Then, for any $a, b \in G$,

$$f(ab) = (ab)H = (aH) \cdot (bH) = f(a) \cdot f(b).$$

Therefore, f is a homomorphism of G into G/H. It is called the natural homomorphism or canonical homomorphism.

Theorem 4.2.54 *Let g be a homomorphism of a group G into a group G'. Let K be the kernel of g and R be the image set of g in G'. Then, G/K is isomorphic to R.*

Proof.

We have already shown that K is a normal subgroup of G. Therefore, there exists a canonical homomorphism $f : G \longrightarrow G/K$ given by $f(a) = aK$, for any $a \in G$.

Now, let us define a mapping $h : G/K \longrightarrow R$ such that $h(aK) = g(a)$.

The image set of h is the same as the image set of g, and hence h is onto. Further, for any $a, b \in G$ such that $aK = bK$, we have $ak_1 = bk_2$, for some $k_1, k_2 \in K$.

Therefore, $g(ak_1) = g(a)g(k_1) = g(a)e' = g(a)$

and $g(bk_2) = g(b)g(k_2) = g(b) = e' = g(b)$

so that $aK = bK \Longrightarrow g(a) = g(b)$.

Therefore, $h(aK) = g(a) = g(b) = h(bK)$.

Also, $f(a) = f(b)$.

Hence, h is one-to-one and onto.

Further, $h(aKbK) = h(abK) == g(ab) = g(a)g(b) = h(aK)h(bK)$.

Hence, h is an isomorphism of G/K to R.

4.2.9 Solved Problems

1. Prove that the intersection of any two subgroups of a group $(G, \star)$ is again a subgroup of $(G, \star)$.

   Solution.

   Let H and K be subgroups of G.

   Let $a \in H \Longrightarrow a, b \in H$ and $a, b \in K$

   $\Longrightarrow \quad a \star b^{-1} \in H$ and $a \star b^{-1} \in K$ (as H and K are subgroups)

   $\Longrightarrow \quad a \star b^{-1} \in H \cap K$.

   Hence, $H \cap K$ is a subgroup of G.

2. Is the union of two subgroups of a group a subgroup of G? Justify your answer.

Solution.

The union of two subgroups of a group G need not be a subgroup of G.

For example, we know $(\mathbb{Z}, +)$ is a group.

Let $H = 3\mathbb{Z} = \{0, \pm 3, \pm 6, \dots\}$

Let $K = 2\mathbb{Z} = \{0, \pm 2, \pm 4, \dots\}$

$\implies$ H and K are subgroups of $(\mathbb{Z}, +)$

$\implies$ $3 \in 3\mathbb{Z} \subseteq 3\mathbb{Z} \cup 2\mathbb{Z} = H \cup K$

$\implies$ $2 \in 2\mathbb{Z} \subseteq 2\mathbb{Z} \cup 3\mathbb{Z} = H \cup K$.

But $3 + 2 = 5 \notin 2\mathbb{Z} \cup 3\mathbb{Z}$.

$\therefore$ $H \cup K$ is not a subgroup of $(\mathbb{Z}, +)$.

3. If H_1 and H_2 are subgroups of a group $(G, \star)$, then prove that $H_1 \cup H_2$ is a subgroup of G if and only if either $H_1 \subseteq H_2$ or $H_2 \subseteq H_1$.

Solution.

Given H_1 and H_2 are two subgroups of $(G, \star)$ and $H_1 \subseteq H_2$ or $H_2 \subseteq H_1$.

If $H_1 \subseteq H_2$, then $H_1 \cup H_2 = H_2$ which is a subgroup of G.

If $H_2 \subseteq H_1$, then $H_1 \cup H_2 = H_1$ which is a subgroup of G.

Conversely, suppose $H_1 \not\subseteq H_2$ and $H_2 \not\subseteq H_1$.

Then, there exist $a \in H_1$ and $a \notin H_2$, and there exist $b \in H_2$ and $b \notin H_1$.

Now $a, b \in H_1 \cup H_2$.

Since $H_1 \cup H_2$ is a subgroup, it follows that $a \star b \in H_1 \cup H_2$.

Hence, $a \star b \in H_1$ or $a \star b \in H_2$.

Case (i): If $a \star b \in H_1$, then $a^{-1} \star (a \star b) \in H_1$.

That is, $b \in H_1$, which is a contradiction.

Case(ii): If $a \star b \in H_2$. Then, $(a \star b) \star b^{-1} \in H_2$.

That is, $a \in H_2$, which is a contradiction.

Thus, either $H_1 \subseteq H_2$ or $H_2 \subseteq H_1$.

4. Find all the subgroups of $(Z_9, +_9)$.

Solution.

$$Z_9 = \{[0], [1], [2], [3], [4], [5], [6], [7], [8]\}.$$

The binary operation is addition modulo 9 (or $+_9$).

Consider the subsets

$$H_1 = \{[0], [2], [4], [6], [8]\}, H_2 = \{[0], [3], [6]\},$$
$$H_3 = \{[0], [4], [8]\}, H_4 = \{[0], [5]\}.$$

The improper subgroups of $(Z_9, +_9)$ are $(\{[0]\}, +_9)$ and $(Z_9, +_9)$.

H_2 is closed since

$+_9$	$[0]$	$[3]$	$[6]$
$[0]$	$[0]$	$[3]$	$[6]$
$[3]$	$[3]$	$[6]$	$[0]$
$[6]$	$[6]$	$[0]$	$[3]$

H_1 is closed since

$+_9$	$[0]$	$[2]$	$[4]$	$[6]$	$[8]$
$[0]$	$[0]$	$[2]$	$[4]$	$[6]$	$[8]$
$[2]$	$[2]$	$[4]$	$[6]$	$[8]$	$[1]$
$[4]$	$[4]$	$[6]$	$[8]$	$[1]$	$[3]$
$[6]$	$[6]$	$[8]$	$[1]$	$[3]$	$[5]$
$[8]$	$[8]$	$[1]$	$[3]$	$[5]$	$[7]$

H_3 is closed since

$+_9$	$[0]$	$[4]$	$[8]$
$[0]$	$[0]$	$[4]$	$[8]$
$[4]$	$[4]$	$[8]$	$[3]$
$[8]$	$[8]$	$[3]$	$[7]$

H_4 is closed since

$+_9$	$[0]$	$[5]$
$[0]$	$[0]$	$[5]$
$[5]$	$[5]$	$[1]$

The above composition tables show that H_1, H_2, H_3, and H_4 are closed under $+_9$. Therefore, the possible subgroups of $(Z_9, +_9)$ are $(H_1, +_9)$, $(H_2, +_9)$, $(H_3, +_9)$, and $(H_4, +_9)$.

5. Find the left cosets of $\{[0], [3]\}$ in the addition modular group $(Z_6, +_6)$.

Solution.

Let $Z_6 = \{[0], [1], [2], [3], [4], [5]\}$ be a group and $H = \{[0], [3]\}$ be a subgroup of Z_6 under $+_6$ (addition mod 6).

The left cosets of H are

$$[0] +_6 H = \{[0], [3]\} = H$$
$$[1] +_6 H = \{[1], [4]\}$$
$$[2] +_6 H = \{[2], [5]\}$$
$$[3] +_6 H = \{[3], [6]\} = \{[3], [0]\} = \{[0], [3]\} = H$$
$$[4] +_6 H = \{[4], [7]\} = \{[4], [1]\} = [1] +_6 H$$
$$[5] +_6 H = \{[5], [8]\} = \{[5], [2]\} = [2] +_6 H.$$

$\therefore \quad [0] +_6 H = [3] +_6 H = H$

and $\quad [1] +_6 H = [4] +_6 H,\ [2] +_6 H = [5] +_6 H$

are the distinct left cosets of H in Z_6.

6. Find the left cosets of $\{[0], [2]\}$ in the group $(Z_4, +_4)$.

   Solution.
   Let $Z_4 = \{[0], [1], [2], [3]\}$ be a group and $H = \{[0], [2]\}$ be a subgroup of Z_4 under $+_4$ (addition mod 4).
   The left cosets of H are

   $$[0] + H = \{[0], [2]\} = H$$
   $$[1] + H = \{[1], [3]\}$$
   $$[2] + H = \{[2], [4]\} = \{[2], [0]\} = \{[0], [2]\} = H$$
   $$[3] + H = \{[3], [5]\} = \{[3], [1]\} = \{[1], [3]\} = [1] + H.$$

   Therefore, $[0] + H = [2] + H = H$ and $[1] + H = [3] + H$ are the two distinct left cosets of H in Z_4.

7. Let $G = \{1, a, a^2, a^3\}$ $(a^4 = 1)$ be a group and $H = \{1, a^2\}$ be a subgroup of G under multiplication. Find all the cosets of H.

   Solution.
   The right cosets of H in G are
   $$H1 = \{a, a^2\} = H$$
   $$Ha = \{a, a^3\}$$
   $$Ha^2 = \{a^2, a^4\} = \{a^2, a\} = H$$
   and $\;\;Ha^3 = \{a^3, a^5\} = \{a^3, a\} = Ha.$

   Therefore, $H1 = H = Ha^2 = \{1, a^2\}$ and $Ha = Ha^3 = \{a, a^3\}$ are two distinct right cosets of H in G. Similarly, we can find the left cosets of H in G.

8. Prove that any two infinite cyclic groups are isomorphic to each other.

   Solution.
   Let $G_1 = <a>$ and $G_2 = <b>$ be two cyclic groups of infinite order.
   $G_1 = \{a^n | n$ is an integer$\}$ and $G_2 = \{b^n | n$ is an integer$\}$.
   Define a map $f : G_1 \longrightarrow G_2$ by $f(a^n) = b^n$.
   Let $x, y \in G_1$. Then, $x = a^n$, $y = a^m$ for some integers n and m. Now,

   $$f(x)f(y) = f(a^n)f(a^m) = b^n b^m = b^{n+m} = f(a^{n+m})$$
   $$= f(a^n a^m) = f(xy).$$

   Hence, f is a homomorphism.
   If $f(x) = f(y)$, then $f(a^n) = f(a^m) \Longrightarrow b^n = b^m$.

   Then, $b^{n-m} = e'$ in G_2. As G_2 is an infinite cyclic group generated by b, there is no non-zero integer k such that $b^k = e'$. Hence, from $b^{n-m} = e'$, we have $n - m = 0$, or $n = m$, and hence $x = a^n = a^m = y$. Thus, f is one-to-one.

Let $z \in G_2$. Then, $z = b^n$ for some integer n. Now, take $x = a^n$. Then, $f(x) = f(a^n) = b^n = z$. So, the map f is onto.

Now, f is one-to-one and onto homomorphism. Hence, it is an isomorphism.

9. Prove that the group homomorphism preserves the identity element.

   Solution.

   Let $(G, \star)$ and $(H, \circ)$ be two groups. Let $f : G \longrightarrow H$ be a group homomorphism.

   Let e_G be the identity element of G.

   Let e_H be the identity element of H.

   To prove: $f(e_G) = e_H$.

   Consider

   $$f(e_G) = f(e_G \star e_G) \quad \text{(since } e_G \text{ is the identity in } G\text{)}$$
   $$= f(e_G) \circ f(e_G) \quad \text{(since } f \text{ is a homomorphism)}$$

   Multiplying bothsides by $f(e_G)^{-1}$ on the right side, we get

   $$f(e_G) \circ f(e_G)^{-1} = f(e_G) \circ f(e_G) \circ f(e_G)^{-1}$$
   $$e_H = f(e_G).$$

   Hence, the group homomorphism preserves the identity element.

10. Prove that the group homomorphism preserves the inverse element.

    Solution.

    Let $(G, \star)$ and $(H, \circ)$ be two groups. Let $f : G \longrightarrow H$ be a group homomorphism.

    Let e_G be the identity element of G.

    Let e_H be the identity element of H.

    To Prove: $[f(x)]^{-1} = f(x^{-1})$, for all $x \in G$.

    It is sufficient to prove that $f(x) \circ f(x^{-1}) = e_H$.

    Now, for all $x \in G$, we can write

    $$f(x) \circ f(x^{-1}) = f(x \star x^{-1}) \text{ (since } f \text{ is a homomorphism)}$$
    $$\implies f(x) \circ f(x^{-1}) = f(e_G) = e_H.$$

    Hence, the group homomorphism preserves the inverse element.

11. If $f : G \longrightarrow G'$ is a group homomorphism from $(G, \star)$ to (G', Δ), then prove that for any $a \in G$, $f(a^{-1}) = [f(a)]^{-1}$.

    Solution.

    For all $a, a^{-1} \in G$, we have

    $$f(a \star a^{-1}) = f(a) \Delta f(a^{-1})$$
    $$\text{or} \qquad f(e) = f(a) \Delta f(a^{-1})$$
    $$\text{or} \qquad e' = f(a) \Delta f(a^{-1}). \tag{4.15}$$

Similarly,

$$f(a^{-1}\Delta a) = f(a^{-1})\Delta f(a)$$

or $$\qquad f(e) = f(a^{-1})\Delta f(a)$$

or $$\qquad e' = f(a^{-1})\Delta f(a). \qquad (4.16)$$

From (4.15) and (4.16), we get

$$f(a)\Delta f(a^{-1}) = f(a^{-1})\Delta f(a)$$

$$\implies \quad f(a^{-1}) \text{ is the inverse of } f(a).$$

That is, $f(a^{-1}) = [f(a)]^{-1}$.

12. Let G be a group and $a \in G$. Let $f : G \longrightarrow G$ be given by $f(x) = axa^{-1}$, for all $x \in G$. Prove that f is an isomorphism of G onto G.

Solution.
To show f is a homomorphism:
If $x, y \in G$, then

$$\begin{aligned}
f(x)f(y) &= (axa^{-1})(aya^{-1}) \\
&= ax(^{-1}a)ya^{-1} \\
&= axya^{-1} \\
&= a(xy)a^{-1} \\
&= f(xy).
\end{aligned}$$

Therefore, f is a homomorphism.

To show f is one-to-one:
If $f(x) = f(y)$, then $axa^{-1} = aya^{-1}$. Hence, by left cancellation law, we have $xa^{-1} = ya^{-1}$; again by right cancellation law, we get $x = y$. Therefore, $f(x) = f(y) \implies x = y$. Hence, f is one-to-one.

To show f is onto:
Let $y \in G$; then $a^{-1}ya \in G$ and

$$\begin{aligned}
f(a^{-1}ya) &= a(a^{-1}ya)a^{-1} \\
&= (aa^{-1})y(aa^{-1}) \\
&= y.
\end{aligned}$$

Therefore, $f(x) = y$, for some $x \in G$.
Thus, f is an isomorphism.

13. Prove that the intersection of two normal subgroups is a normal subgroup.

Solution.
Let H and K be any two normal subgroups of a group G. We have to prove that $H \cap K$ is a normal subgroup of G.

Since H and K are subgroups of G, $e \in H$ and $e \in H$. Hence, $e \in H \cap K$. Thus, $H \cap K$ is a non-empty set.

Let $a, b \in H \cap K$.

Claim: $ab^{-1} \in H \cap K$.

Since $a, b \in H \cap K$, both a and b belong to H and K.

Since H and K are subgroups of G, $ab^{-1} \in H$ and $ab^{-1} \in K$, so that $ab^{-1} \in H \cap K$.

Hence, $H \cap K$ is a subgroup of G, by a criterion for subgroup.

To prove: $H \cap K$ is normal.

Let $x \in H \cap K$ and $g \in H$.

Since $x \in H \cap K$, $x \in H$ and $x \in K$.

Since $x \in H$, $g \in G \Longrightarrow gxg^{-1} \in K$ (as H is normal).

Similarly, $x \in K$, $g \in G \Longrightarrow gxg^{-1} \in K$ (as K is normal).

Hence, $x \in H \cap K$ and $g \in G \Longrightarrow gxg^{-1} \in H \cap K$.

Thus, $H \cap K$ is a normal subgroup of G.

14. Prove that every subgroup of an abelian group is a normal subgroup.

Solution.

Let $(G, \star)$ be an abelian group and $(N, \star)$ be a subgroup of G.

Let g be any element in G, and let $n \in N$.

Now,

$$g \star n \star g^{-1} = (n \star g) \star g^{-1} \quad [\because \ G \text{ is abelian}]$$
$$= n \star (g \star g^{-1})$$
$$= n \star e$$
$$= n \in N.$$

Therefore, for all $g \in G$ and $n \in N$, $g \star n \star g^{-1} \in N$.

Hence, $(N, \star)$ is a normal subgroup.

4.2.10   Permutation Functions

Definition 4.2.55 *A bijection from a set A to itself is called a permutation of A.*

Example 4.2.56 *Let $A = \mathbb{R}$, and let $f : A \longrightarrow A$ be defined by $f(a) = 2a + 1$. Since f is one-to-one and onto, it follows that f is a permutation of A.*

Example 4.2.57 *Let $A = \{1, 2, 3\}$. Then, all the permutations of A are*

$$1_A = \begin{pmatrix} 1 & 2 & 3 \\ 1 & 2 & 3 \end{pmatrix}, p_1 = \begin{pmatrix} 1 & 2 & 3 \\ 1 & 3 & 2 \end{pmatrix}, p_2 = \begin{pmatrix} 1 & 2 & 3 \\ 2 & 1 & 3 \end{pmatrix},$$

$$p_3 = \begin{pmatrix} 1 & 2 & 3 \\ 2 & 3 & 1 \end{pmatrix}, p_4 = \begin{pmatrix} 1 & 2 & 3 \\ 3 & 1 & 2 \end{pmatrix}, p_5 = \begin{pmatrix} 1 & 2 & 3 \\ 3 & 2 & 1 \end{pmatrix}.$$

Remark 4.2.58 *In the above example, we can write the permutations as ordered pairs. For example,*

$$p_4 = \{(1,3),(2,1),(3,2)\} \quad and \quad p_4^{-1} = \{(3,1),(1,2),(2,3)\}.$$

Or, if the first component of each ordered pair is written in increasing order, then we have

$$p_4^{-1} = \{(1,2),(2,3),(3,1)\}.$$

Thus, $\quad p_4^{-1} = \begin{pmatrix} 1 & 2 & 3 \\ 2 & 3 & 1 \end{pmatrix} = p_3.$

Remark 4.2.59 *The function p_2 takes 1 to 2, and p_3 takes 2 to 3, so $p_3 \circ p_2$ takes 1 to 3. Also, p_2 takes 2 to 1, and p_3 takes 1 to 2, so $p_3 \circ p_2$ takes 2 to 2. Finally, p_2 takes 3 to 3, and p_3 takes 3 to 1, so $p_3 \circ p_2$ takes 3 to 1. Thus,*

$$p_3 \circ p_2 = \begin{pmatrix} 1 & 2 & 3 \\ 3 & 2 & 1 \end{pmatrix}.$$

Remark 4.2.60 *The process of forming $p_3 \circ p_2$ is shown below. It can be noted that $p_3 \circ p_2 = p_5$.*

$$p_3 \circ p_2 = \begin{pmatrix} \overleftarrow{} \\ 1 & 2 & 3 \\ \downarrow \\ 2 & 3 & 1 \end{pmatrix} \uparrow \circ \begin{pmatrix} 1 & 2 & 3 \\ \downarrow \\ 2 & 1 & 3 \\ \overleftarrow{} \end{pmatrix} = \begin{pmatrix} 1 & 2 & 3 \\ 3 & 2 & 1 \end{pmatrix} = p_5$$

Theorem 4.2.61 *If $A = \{a_1, a_2, \ldots, a_n\}$ is a set containing n elements, then there are $n! = n \cdot (n-1) \cdots 2 \cdot 1$ permutations of A.*

Definition 4.2.62 Cyclic Permutation: *Let $b_1, b_2, \ldots, b_r$ be r distinct elements of the set $A = \{a_1, a_2, \ldots, a_n\}$. The permutation $p : A \longrightarrow A$ is defined by*

$$p(b_1) = b_2$$
$$p(b-2) = b_3$$

$$\vdots \quad \vdots$$

$$p(b_{r-1}) = b_r$$
$$p(b_r) = b_1.$$

$p(x) = x$ if $x \in A$, $x \notin \{b_1, b_2, \ldots, b_r\}$ is called a cyclic permutation of length r, or simply a cycle of length of r, and will be denoted by $(b_1, b_2, \ldots, b_r)$.

Example 4.2.63 *Let $A = \{1, 2, 3, 4, 5\}$. The cycle $(1, 3, 5)$ denotes the permutation*

$$\begin{pmatrix} 1 & 2 & 3 & 4 & 5 \\ 3 & 2 & 5 & 4 & 1 \end{pmatrix}.$$

Definition 4.2.64 Disjoint Cycles: *Two cycles of a set A are said to be disjoint if no element of A appears in both cycles.*

Example 4.2.65 *Let* $A = \{1, 2, 3, 4, 5, 6\}$. *Then, the cycles* $(1, 2, 5)$ *and* $(3, 4, 6)$ *are disjoint, whereas the cycles* $(1, 2, 5)$ *and* $(2, 4, 6)$ *are not.*

Theorem 4.2.66 *A permutation of a finite set that is not the identity or a cycle can be written as a product of disjoint cycles of length* ≥ 2.

Definition 4.2.67 Transposition: *A cycle of length 2 is called a transposition. That is, a transposition is a cycle* $p = (a_i, a_j)$, *where* $p(a_i) = a_j$ *and* $p(a_j) = a_i$.

Remark 4.2.68 *Note that if* $p = (a_i, a_j)$ *is a transposition of* A, *then* $p \circ p = 1_A$, *the identity permutation of* A.
     Every cycle can be written as a product of transpositions. In fact,
$$(b_1, b_2, \ldots, b_r) = (b_1, b_r) \circ (b_1, b_{r-1}) \circ \cdots \circ (b_1, b_3) \circ (b_1, b_2).$$
This case can be verified by induction on r *as follows.*

Basis Step:
If $r = 2$, *then the cycle is just* (b_1, b_2), *which already has the proper form.*

Induction Step:
We use $P(m)$ *to find* $P(m+1)$. *Let* $(b_1, b_2, \ldots, b_m, b_{m+1})$ *be a cycle of length* $m + 1$. *Then,* $(b-1, b-2, \ldots, b_m, b_{m+1}) = (b_1, b_{m+1}) \circ (b_1, b_2, \ldots, b_m)$ *as may be verified by computing the composition.*
     Using $P(m)$, $(b_1, b_2, \ldots, b_m) = (b_1, b_k) \circ (b_1, b_{m-1}) \circ \cdots \circ (b_1, b_2)$.
     Thus, by substitution,
$$(b_1, b_2, \ldots, b_{m+1}) = (b_1, b_{m+1}) \circ (b_1, b_m) \circ \cdots \circ (b_1, b_3) \circ (b_1, b_2).$$
This completes the induction step. Thus, by the principle of mathematical induction, the result holds for every cycle. For example,
$$(1, 2, 3, 4, 5) = (1, 5) \circ (1, 4) \circ (1, 3) \circ (1, 2).$$

Corollary 4.2.69 *Every permutation of a finite set with at least two elements can be written as a product of transpositions.*

Theorem 4.2.70 *If a permutation of a finite set can be written as a product of an even number of transpositions, then it can never be written as a product of an odd number of transpositions, and the converse is also true.*

Definition 4.2.71 Even and Odd Permutations: *A permutation of a finite set is called even if it can be written as a product of an even number of transpositions, and it is called odd if it can be written as a product of an odd number of transpositions.*

Remark 4.2.72 *From the definition of even and odd permutations, we have the following:*

    (a) *The product of two even permutations is even.*

    (b) *The product of two odd permutations is even.*

    (c) *The product of an even and an odd permutation is odd.*

Theorem 4.2.73 *Let $A = \{a_1, a_2, \ldots, a_n\}$ be a finite set with n elements, $n \geq 2$. Then, there are $\dfrac{n!}{2}$ even permutations and $\dfrac{n!}{2}$ odd permutations.*

Proof.

Let A_n be the set of all even permutations of A, and let B_n be the set of all odd permutations. We shall define a function $f : A_n \longrightarrow B_n$, which we shall show is one-to-one and onto, and this will show that A_n and B_n have the same number of elements.

Since $n \geq 2$, we can choose a particular transposition q_0 of A, say $q_0 = (a_{n-1}, a_n)$. We now define the function $f : A_n \longrightarrow B_n$ by

$$f(p) = q_0 \circ p, \quad p \in A_n.$$

Observe that if $p \in A_n$, then p is an even permutation, so $q_0 \circ p$ is an odd permutation, and thus $f(p) \in B_n$.

Suppose now that p_1 and p_2 are in A_n and $f(p_1) = f(p_2)$. Then

$$q_0 \circ p_1 = q_0 \circ p_2. \tag{4.17}$$

We now compose each side of equation (4.17) with q_0:

$$q_0 \circ (q_0 \circ p_1) = q_0 \circ (q_0 \circ p_2);$$

so by the associative property,

$$(q_0 \circ q_0) \circ p_1 = (q_0 \circ q_0) \circ p_2, \text{ or since } q_0 \circ q_0 = 1_A,$$
$$1_A \circ p_1 = 1_A \circ p_2$$
$$p_1 = p_2.$$

Thus, f is one-to-one.

Now, let $q \in B_n$. Then, $q_0 \circ q \in A_n$, and

$f(q_0 \circ q) = q_0 \circ (q_0 \circ q) = (q_0 \circ q_0) = 1_A \circ q = q,$

which means that f is an onto function. Since $f : A_n \longrightarrow B_n$ is one-to-one and onto, we conclude that A_n and B_n have the same number of elements. Note that $A_n \cap B_n = \phi$, since no permutation can be both even and odd. Also, by theorem, $|A_n \cup B_n| = n!$.

$n! = |A_n \cup B_n| = |A_n| + |B_n| - |A_n \cap B_n| = 2|A_n|.$

Hence, we have $|A_n| = |B_n| = \dfrac{n!}{2}.$

4.2.11 Solved Problems

1. Let $A = \{1, 2, 3, 4, 5, 6\}$. Compute $(4, 1, 3, 5) \circ (5, 6, 3)$ and $(5, 6, 3) \circ (4, 1, 3, 5)$.

Solution.
We have

$$(4,1,3,5) = \begin{pmatrix} 1 & 2 & 3 & 4 & 5 & 6 \\ 3 & 2 & 5 & 1 & 4 & 6 \end{pmatrix} \quad \text{and} \quad (5,6,3) = \begin{pmatrix} 1 & 2 & 3 & 4 & 5 & 6 \\ 1 & 2 & 5 & 4 & 6 & 3 \end{pmatrix}.$$

Then, $(4,1,3,5) \circ (5,6,3) = \begin{pmatrix} 1 & 2 & 3 & 4 & 5 & 6 \\ 3 & 2 & 5 & 1 & 4 & 6 \end{pmatrix} \circ \begin{pmatrix} 1 & 2 & 3 & 4 & 5 & 6 \\ 1 & 2 & 5 & 4 & 6 & 3 \end{pmatrix}$

$$= \begin{pmatrix} 1 & 2 & 3 & 4 & 5 & 6 \\ 3 & 2 & 4 & 1 & 6 & 5 \end{pmatrix}$$

and $(5,6,3) \circ (4,1,3,5) = \begin{pmatrix} 1 & 2 & 3 & 4 & 5 & 6 \\ 1 & 2 & 5 & 4 & 6 & 3 \end{pmatrix} \circ \begin{pmatrix} 1 & 2 & 3 & 4 & 5 & 6 \\ 3 & 2 & 5 & 1 & 4 & 6 \end{pmatrix}$

$$= \begin{pmatrix} 1 & 2 & 3 & 4 & 5 & 6 \\ 5 & 2 & 6 & 1 & 4 & 3 \end{pmatrix}.$$

Observe that

$$(4,1,3,5) \circ (5,6,3) \neq (5,6,3) \circ (4,1,3,5)$$

and that neither product is a cycle.

2. Let $A = \{1,2,3,4,5,6,7,8\}$ be a set. Then, write the permutation
$p = \begin{pmatrix} 1 & 2 & 3 & 4 & 5 & 6 & 7 & 8 \\ 3 & 4 & 6 & 5 & 2 & 1 & 8 & 7 \end{pmatrix}$ as a product of disjoint cycles.

Solution.
We start with 1 and find that $p(1) = 3$, $p(3) = 6$, and $p(6) = 1$, so we have the cycle $(1,3,6)$. Next, we choose the first element of A that has not appeared in a previous cycle. We choose 2, and we have $P(2) = 4$, $p(4) = 5$, and $p(5) = 2$, so we obtain the cycle $(2,4,5)$. We now choose 7, the first element of A that has not appeared in a previous cycle. Since $p(7) = 8$ and $p(8) = 7$, we obtain the cycle $(7,8)$. We can then write p as a product of disjoint cycles as

$$p = (7,8) \circ (2,4,5) \circ (1,3,6).$$

3. Is the permutation $p = \begin{pmatrix} 1 & 2 & 3 & 4 & 5 & 6 & 7 \\ 2 & 4 & 5 & 7 & 6 & 3 & 1 \end{pmatrix}$ even or odd?

Solution.
We first write p as a product of disjoint cycles, obtaining

$$p = (3,5,6) \circ (1,2,4,7).$$

Next, we write each of the cycles as a product of transpositions:

$$(1,2,4,7) = (1,7) \circ (1,4) \circ (1,2)$$
$$(3,5,6) = (3,6) \circ (3,5).$$

Then, $p = (3,6) \circ (3,5) \circ (1,7) \circ (1,4) \circ (1,2)$. Since p is a product of an odd number of transpositions, it is an odd permutation.

4. Show that the permutation $\begin{pmatrix} 1 & 2 & 3 & 4 & 5 & 6 \\ 5 & 6 & 2 & 4 & 1 & 3 \end{pmatrix}$ is odd, while the permutation $\begin{pmatrix} 1 & 2 & 3 & 4 & 5 & 6 \\ 6 & 3 & 4 & 5 & 2 & 1 \end{pmatrix}$ is even.

Solution.

$$\begin{pmatrix} 1 & 2 & 3 & 4 & 5 & 6 \\ 5 & 6 & 2 & 4 & 1 & 3 \end{pmatrix} = (1\ \ 5)(2\ \ 6\ \ 3) = (1\ \ 5)(2\ \ 6)(2\ \ 3).$$

The given permutation can be expressed as the product of an odd number of transpositions, and hence the permutation is odd. Again

$$\begin{pmatrix} 1 & 2 & 3 & 4 & 5 & 6 \\ 6 & 3 & 4 & 5 & 2 & 1 \end{pmatrix} = (1\ \ 6)(2\ \ 3\ \ 4\ \ 5) = (1\ \ 6)(2\ \ 3)(2\ \ 4)(2\ \ 5).$$

Since it is a product of even number of transpositions, the permutation is an even permutation.

5. Express the permutation $\begin{pmatrix} 1 & 2 & 3 & 4 & 5 & 6 \\ 6 & 5 & 2 & 4 & 3 & 1 \end{pmatrix}$ as a product of transpositions.

Solution.

$$\begin{pmatrix} 1 & 2 & 3 & 4 & 5 & 6 \\ 6 & 5 & 2 & 4 & 3 & 1 \end{pmatrix} = (1\ \ 6)(2\ \ 5\ \ 3) = (1\ \ 6)(2\ \ 5)(2\ \ 3).$$

6. Find the inverse of the permutation $\begin{pmatrix} 1 & 2 & 3 & 4 & 5 \\ 2 & 3 & 1 & 5 & 4 \end{pmatrix}$.

Solution.

$$\text{Given } \begin{pmatrix} 1 & 2 & 3 & 4 & 5 \\ 2 & 3 & 1 & 5 & 4 \end{pmatrix}.$$

Let the inverse permutation be $\begin{pmatrix} 1 & 2 & 3 & 4 & 5 \\ x & y & z & u & v \end{pmatrix}$.

Then, $\begin{pmatrix} 1 & 2 & 3 & 4 & 5 \\ 2 & 3 & 1 & 5 & 4 \end{pmatrix} \begin{pmatrix} 1 & 2 & 3 & 4 & 5 \\ x & y & z & u & v \end{pmatrix} = \begin{pmatrix} 1 & 2 & 3 & 4 & 5 \\ 1 & 2 & 3 & 4 & 5 \end{pmatrix}$

$\implies \begin{pmatrix} 1 & 2 & 3 & 4 & 5 \\ y & z & x & v & u \end{pmatrix} = \begin{pmatrix} 1 & 2 & 3 & 4 & 5 \\ 1 & 2 & 3 & 4 & 5 \end{pmatrix}$

$\implies y = 1, z = 2, x = 3, v = 4, u = 5.$

Hence, the inverse permutation is $\begin{pmatrix} 1 & 2 & 3 & 4 & 5 \\ 3 & 1 & 2 & 5 & 4 \end{pmatrix}$.

7. If $A = (1\ \ 2\ \ 3\ \ 4\ \ 5)$, $B = (2\ \ 3)(4\ \ 5)$, find AB.

Solution.

Given $A = (1\ 2\ 3\ 4\ 5)$, $B = (2\ 3)(4\ 5)$.

$$AB = \begin{pmatrix} 1 & 2 & 3 & 4 & 5 \\ 2 & 3 & 4 & 5 & 1 \end{pmatrix} \begin{pmatrix} 1 & 2 & 3 & 4 & 5 \\ 1 & 3 & 2 & 5 & 4 \end{pmatrix}$$

$$= \begin{pmatrix} 1 & 2 & 3 & 4 & 5 \\ 3 & 2 & 5 & 4 & 1 \end{pmatrix}$$

$$= (1\ 3\ 5).$$

8. If $A = \{1, 2, 3, 4, 5, 6, 7, 8\}$, then express the following permutations as a product of disjoint cycles.

   (i) $p = \begin{pmatrix} 1 & 2 & 3 & 4 & 5 & 6 & 7 & 8 \\ 6 & 5 & 7 & 8 & 4 & 3 & 2 & 1 \end{pmatrix}$

   (ii) $p = \begin{pmatrix} 1 & 2 & 3 & 4 & 5 & 6 & 7 & 8 \\ 2 & 3 & 1 & 4 & 6 & 7 & 8 & 5 \end{pmatrix}$.

Solution.

   (i) $p(1) = 6, p(6) = 3, p(3) = 7, p(7) = 2, p(2) = 5, p(5) = 4$,
       $p(4) = 8, p(8) = 1$.
          Therefore, $p = (1\ 6\ 3\ 7\ 2\ 5\ 4\ 8)$.

   (ii) $p(1) = 2, p(2) = 3, p(3) = 1 \Longrightarrow (1\ 2\ 3)$
        $p(5) = 6, p(6) = 7, p(7) = 8, p(8) = 5 \Longrightarrow (5\ 6\ 7\ 8)$.
           Therefore, $p = (5\ 6\ 7\ 8)(1\ 2\ 3)$.

9. Let $A = \{1, 2, 3, 4, 5, 6\}$ and $p = \begin{pmatrix} 1 & 2 & 3 & 4 & 5 & 6 \\ 2 & 4 & 3 & 1 & 5 & 6 \end{pmatrix}$ be a permutation of A.

   (i) Write p as a product of disjoint cycles.
   (ii) Compute p^{-1}.
   (iii) Compute p^2.
   (iv) Find the period of p, that is, the smallest positive integer k such that $p^k = 1_A$.

Solution.

   (i) Given $p = \begin{pmatrix} 1 & 2 & 3 & 4 & 5 & 6 \\ 2 & 4 & 3 & 1 & 5 & 6 \end{pmatrix}$.

      Since $p(1) = 2, p(2) = 4$, and $p(4) = 1$, we write $p = (1, 2, 4)$ as the other elements are fixed.

   (ii) $p^{-1} = \begin{pmatrix} 2 & 4 & 3 & 1 & 5 & 6 \\ 1 & 2 & 3 & 4 & 5 & 6 \end{pmatrix} = p = \begin{pmatrix} 1 & 2 & 3 & 4 & 5 & 6 \\ 4 & 1 & 3 & 2 & 5 & 6 \end{pmatrix}$.

   (iii) $p^2 = p \circ p = \begin{pmatrix} 1 & 2 & 3 & 4 & 5 & 6 \\ 4 & 1 & 3 & 2 & 5 & 6 \end{pmatrix}$.

(iv) $p^3 = p^2 \circ p = \begin{pmatrix} 1 & 2 & 3 & 4 & 5 & 6 \\ 1 & 2 & 3 & 4 & 5 & 6 \end{pmatrix} = 1_A.$

$p^4 = p, p^5 = p^2,$ etc.

Therefore, the period of $p = 3$.

10. If $f = \begin{pmatrix} 1 & 2 & 3 & 4 \\ 3 & 2 & 1 & 4 \end{pmatrix}$ and $g = \begin{pmatrix} 1 & 2 & 3 & 4 \\ 2 & 3 & 4 & 1 \end{pmatrix}$ are permutations, prove that $(g \circ f)^{-1} = f^{-1} \circ g^{-1}$.

Solution.

$$f^{-1} = \begin{pmatrix} 3 & 2 & 1 & 4 \\ 1 & 2 & 3 & 4 \end{pmatrix} \text{ and } g^{-1} = \begin{pmatrix} 1 & 2 & 3 & 4 \\ 4 & 1 & 2 & 3 \end{pmatrix}.$$

$$f^{-1} \circ g^{-1} = \begin{pmatrix} 1 & 2 & 3 & 4 \\ 4 & 3 & 2 & 1 \end{pmatrix}$$

$$g \circ f = \begin{pmatrix} 1 & 2 & 3 & 4 \\ 2 & 3 & 4 & 1 \end{pmatrix} \circ \begin{pmatrix} 1 & 2 & 3 & 4 \\ 3 & 2 & 1 & 4 \end{pmatrix} = \begin{pmatrix} 1 & 2 & 3 & 4 \\ 4 & 3 & 2 & 1 \end{pmatrix}$$

$$(g \circ f)^{-1} = \begin{pmatrix} 1 & 2 & 3 & 4 \\ 4 & 3 & 2 & 1 \end{pmatrix}.$$

Hence, $(g \circ f)^{-1} = f^{-1} \circ g^{-1}$.

11. Let $p_1 = \begin{pmatrix} 1 & 2 & 3 & 4 & 5 & 6 & 7 \\ 7 & 3 & 2 & 1 & 4 & 5 & 6 \end{pmatrix}$ and $p_2 = \begin{pmatrix} 1 & 2 & 3 & 4 & 5 & 6 & 7 \\ 6 & 3 & 2 & 1 & 5 & 4 & 7 \end{pmatrix}$.

(i) Compute $p_1 \circ p_2$.

(ii) Compute p_1^{-1}.

(iii) Is p_1 an even or odd permutation? Explain.

Solution.

(i) $p_1 \circ p_2 = \begin{pmatrix} 1 & 2 & 3 & 4 & 5 & 6 & 7 \\ 7 & 3 & 2 & 1 & 4 & 5 & 6 \end{pmatrix} \circ \begin{pmatrix} 1 & 2 & 3 & 4 & 5 & 6 & 7 \\ 6 & 3 & 2 & 1 & 5 & 4 & 7 \end{pmatrix}$

$= \begin{pmatrix} 1 & 2 & 3 & 4 & 5 & 6 & 7 \\ 5 & 2 & 3 & 7 & 4 & 1 & 6 \end{pmatrix}.$

(ii) $p_1^{-1} = \begin{pmatrix} 1 & 2 & 3 & 4 & 5 & 6 & 7 \\ 4 & 3 & 2 & 5 & 6 & 7 & 1 \end{pmatrix}.$

(iii) $p_1 = (1 \ 7 \ 6 \ 5 \ 4) \circ (2 \ 3)$

$= (1 \ 4) \circ (1 \ 5) \circ (1 \ 6) \circ (1 \ 7) \circ (2 \ 3)$

$=$ product of odd number of transpositions.

Therefore, p_1 is an odd permutation.

12. If $x = (1 \ 2 \ 3)$, $y = (2 \ 4 \ 3)$, and $z = (1 \ 3 \ 4)$, then show that $xyz = 1$.

Solution.

Given $x = (1\ 2\ 3) = \begin{pmatrix} 1 & 2 & 3 & 4 \\ 2 & 3 & 1 & 4 \end{pmatrix}$

$$y = (2\ 4\ 3) = \begin{pmatrix} 1 & 2 & 3 & 4 \\ 1 & 4 & 2 & 3 \end{pmatrix}$$

$$z = (1\ 3\ 4) = \begin{pmatrix} 1 & 2 & 3 & 4 \\ 3 & 2 & 4 & 1 \end{pmatrix}.$$

$$\text{Therefore, } xyz = \begin{pmatrix} 1 & 2 & 3 & 4 \\ 2 & 3 & 1 & 4 \end{pmatrix} \circ \begin{pmatrix} 1 & 2 & 3 & 4 \\ 1 & 4 & 2 & 3 \end{pmatrix} \circ \begin{pmatrix} 1 & 2 & 3 & 4 \\ 3 & 2 & 4 & 1 \end{pmatrix}$$

$$= \begin{pmatrix} 1 & 2 & 3 & 4 \\ 4 & 2 & 1 & 3 \end{pmatrix} \circ \begin{pmatrix} 1 & 2 & 3 & 4 \\ 3 & 2 & 1 & 4 \end{pmatrix}$$

$$= \begin{pmatrix} 1 & 2 & 3 & 4 \\ 1 & 2 & 3 & 4 \end{pmatrix} = 1.$$

4.2.12   Problems for Practice

1. Which of the following functions $f : \mathbb{Z} \longrightarrow \mathbb{Z}$ are permutations of $\mathbb{Z}$?

   (i) f is defined by $f(a) = a + 1$.
   (ii) f is defined by $f(a) = (a - 1)^2$.

2. Which of the following functions $f : \mathbb{R} \longrightarrow \mathbb{R}$ are permutations of $\mathbb{R}$?

   (i) f is defined by $f(a) = a^3$.
   (ii) f is defined by $f(a) = e^a$.

3. Which of the following functions $f : \mathbb{R} \longrightarrow \mathbb{R}$ are permutations of $\mathbb{R}$?

   (i) f is defined by $f(a) = a - 1$.
   (ii) f is defined by $f(a) = a^2$.

4. Which of the following functions $f : \mathbb{Z} \longrightarrow \mathbb{Z}$ are permutations of $\mathbb{Z}$?

   (i) f is defined by $f(a) = a^2 + 1$.
   (ii) f is defined by $f(a) = a^3 - 3$.

5. Let $A = \{a, b, c, d, e, f, g\}$. Write each of the following permutations as a product of disjoint cycles.

   (i) $\begin{pmatrix} a & b & c & d & e & f & g \\ g & d & b & a & c & f & e \end{pmatrix}$

   (ii) $\begin{pmatrix} a & b & c & d & e & f & g \\ d & e & a & b & g & f & c \end{pmatrix}$

6. Let $A = \{1, 2, 3, 4, 5, 6, 7, 8\}$. Write each of the following permutations as a product of transpositions.

   (i) $(2\ 1\ 4\ 5\ 8\ 6)$
   (ii) $(3\ 1\ 6) \circ (4\ 8\ 2\ 5)$

7. Code the message "**WHERE ARE YOU**" by applying the permutation
$$(1\ 7\ 3\ 5\ 11) \circ (2\ 6\ 9) \circ (4\ 8\ 10).$$

8. Decode the message "**ATEHAOMOMNTI**", which was encoded using the permutation
$$(3\ 7\ 1\ 12) \circ (2\ 5\ 8) \circ (4\ 10\ 6\ 11\ 9).$$

9. Let $A = \{1,2,3,4,5,6,7,8\}$. Determine whether the following permutations are even or odd.

(i) $\begin{pmatrix} 1 & 2 & 3 & 4 & 5 & 6 & 7 & 8 \\ 4 & 2 & 1 & 6 & 5 & 8 & 7 & 3 \end{pmatrix}$

(ii) $\begin{pmatrix} 1 & 2 & 3 & 4 & 5 & 6 & 7 & 8 \\ 7 & 3 & 4 & 2 & 1 & 8 & 6 & 5 \end{pmatrix}$

(iii) $(6\ 4\ 2\ 1\ 5)$

(iv) $(4\ 8) \circ (3\ 5\ 2\ 1) \circ (2\ 4\ 7\ 1)$

10. Prove that the product of two even permutations is even.

11. Prove that the product of two odd permutations is even.

12. Prove that the product of an even and odd permutation is odd.

13. Let $A = \{1,2,3,4,5\}$. Let $f = (5\ 2\ 3)$ and $g = (3\ 4\ 1)$ be permutations of A. Compute each of the following, and write the result as the product of disjoint cycles:

(i) $f \circ g$

(ii) $f^{-1} \circ g^{-1}$.

4.2.13 Rings and Fields

Definition 4.2.74 Ring: *An algebraic system $(S, +\cdot)$ is called a ring if the binary operations $+$ and $\cdot$ on S satisfy the following three properties:*

1. *$(S, +)$ is an abelian group.*

2. *$(S, \cdot)$ is a semigroup.*

3. *The operation $\cdot$ is distributive over $+$; that is, for any $a, b, c \in S$,*
$$a \cdot (b + c) = a \cdot b + a \cdot c \qquad and \qquad (b + c) \cdot a = b \cdot a + c \cdot a.$$

Examples:

1. The set of all integers $\mathbb{Z}$, the set of all rational numbers $\mathbb{Q}$, the set of all real numbers $\mathbb{R}$ are rings under the usual addition and usual multiplication.

2. The set of all $n \times n$ matrices M_n is a ring under the matrix addition and matrix multiplication.

3. If n is a positive integer, then $Z_n = \{[0], [1], \ldots, [n-1]\}$ is a ring under $+_n$, the addition modulo n, and $\times_n$, the multiplication modulo n.

4. Let $(R, +, \cdot)$ be a ring and X be a non-empty set. Let A be the set of all functions from X to R. That is, $A = \{f | f : X \longrightarrow R$ is a function$\}$. We define $\oplus$ and $\cdot$ on A as follows:

   (i) if $f, g \in A$, then $f \oplus g : X \longrightarrow R$ is given by

$$(f \oplus g)(x) = f(x) + g(x), \quad \text{for all} \ \ x \in X.$$

   (ii) if $f, g \in A$, then $f, g : X \longrightarrow R$ is given by

$$(f \cdot g)(x) = f(x) \cdot g(x), \quad \text{for all} \ \ x \in X.$$

Definition 4.2.75 Integral Domain: *A commutative ring* $(S, +, \bullet)$ *with identity and without divisors of zero is called an integral domain.*

Definition 4.2.76 Field: *A commutative ring* $(S, +, \bullet)$ *which has more than one element such that every non-zero element of S has a multiplicative inverse in S is called a field.*

Definition 4.2.77 Subring: *A subset R of a ring* $(S, +, \bullet)$ *is called a subring if* $(R, +, \bullet)$ *itself is a ring with the operations* $+$ *and* $\bullet$ *restricted to R.*

Examples:

1. The set of integers $\mathbb{Z}$ is a subring of the ring of all rational numbers $\mathbb{Q}$.

2. The set of all even integers is a subring of the ring of all integers $\mathbb{Z}$.

Definition 4.2.78 Ring Homomorphism: *Let* $(R, +, \bullet)$ *and* $(S, \oplus, \odot)$ *be rings. A mapping* $g : R \longrightarrow S$ *is called a ring homomorphism from* $(R, +, \bullet)$ *to* $(S, \oplus, \odot)$ *if for any* $a, b \in R,$

$$g(a + b) = g(a) \oplus g(b) \ and$$
$$g(a \cdot b) = g(a) \odot g(b).$$

Examples:

1. The ring M_n of all non-zero matrices is not commutative and has non-zero divisors. For example, let $n = 2$; then if $A = \begin{pmatrix} 0 & 1 \\ 0 & 0 \end{pmatrix}$ and $B = \begin{pmatrix} 1 & 0 \\ 0 & 0 \end{pmatrix}$, then $AB = \begin{pmatrix} 0 & 0 \\ 0 & 0 \end{pmatrix}$ and $BA = \begin{pmatrix} 0 & 1 \\ 0 & 0 \end{pmatrix}$. So, $AB \neq BA$, and A is a non-zero divisor.

2. The ring $\mathbb{Q}$ of rational numbers and the ring $\mathbb{R}$ of real numbers are fields.

3. The ring $(Z_7, +_7, \times_7)$ is a field.

4. The ring $(Z_{10}, +_{10}, \times_{10})$ is not an integral domain since $5 \times_{10} 2 = 0$, even though $5 \neq 0$, $2 \neq 0$ in Z_{10}.

5. The ring $\mathbb{Z}$ of all integers is an integral domain but not a field.

Definition 4.2.79 Commutative Ring: *A ring $(R, +, \cdot)$ is said to be commutative if $a \cdot b = b \cdot a$, for all $a, b \in R$.*

Theorem 4.2.80 *Every finite integral domain is a field.*

Proof.
Let $(R, +, \bullet)$ be a finite integral domain.

To prove: $(R - \{0\}, \bullet)$ is a group, that is, to prove

(i) there exists an element $1 \in R$ such that
$$1 \cdot a = a \cdot 1 = a, \text{ for all } a \in R \quad \text{(since } 1 \in R \text{ is an identity)}$$

(ii) for every element of $0 \neq a \in R$, there exists an element $a^{-1} \in R$ such that
$$a \cdot a^{-1} = a^{-1} \cdot a = a.$$

Let $R - \{0\} = \{a_1, a_2, a_3, \ldots, a_n\}$.

Let $a \in R - \{0\}$. Then, the elements $aa_1, aa_2, \ldots, aa_n$ are all in $R - \{0\}$, and they are all distinct. That is, if $a \cdot a_i = a \cdot a_j$, $i \neq j$, then $a \cdot (a_i - a_j) = 0$.

Since R is an integral domain and $a \neq 0$, we must have
$$a_i - a_j = 0 \implies a_i = a_j, \text{ which is a contradiction.}$$

Therefore, $R - \{0\}$ has exactly n elements, and R is a commutative ring with cancellation law. Hence, we get

$a = a \cdot a_{i_0}$, for some i_0 (since $a \in R - \{0\}$).

That is, $a \cdot a_{i_0} = a_{i_0} \cdot a$ (since R is commutative).

Thus, let $x = a \cdot a_i$ for some $a_i \in R - \{0\}$, and
$$y \cdot a_{i_0} = a \cdot a_{i_0} = (a_i \cdot a)a_{i_0} = a_i \cdot a = a \cdot a_j = y.$$

Therefore, a_{i_0} is unity in $R - \{0\}$. We write it as 1.

Since $1 \in R - \{0\}$, there exists an element $aa_k \in R - \{0\}$ such that

$$aa_k = 1.$$

Therefore, $ba = b = 1$ (let $a_k = b$).

Hence, b is the inverse of a, and the converse is also true.

Hence, $(R, +, \bullet)$ is a field.

Theorem 4.2.81 *Every field is an integral domain.*

Proof.
Let $(F, +, \cdot)$ be a field. That is, F is a commutative ring with unity.

To prove F is an integral domain, it is enough to show that it has no zero divisor.

Let $a, b \in F$ such that $a \cdot b = 0$.

If $a \neq 0$, then $a^{-1} \in F$.

Therefore, $a \cdot b = 0$

$\implies \quad a^{-1} \cdot (a \cdot b) = a^{-1} \cdot 0$

$\implies \quad 1 \cdot b = 0$

$\implies \quad b = 0.$

Hence, the theorem is proved.

Note:

The converse of the above Theorem 4.2.81 need not be true.

Theorem 4.2.82 *A commutative ring* $(R, +, \cdot)$ *is an integral domain if and only if the cancellation law holds in* R. *That is,*

$$\text{for } a \neq 0, \quad a \cdot b = a \cdot c \implies b = c, \quad \text{for all} \quad a, b, c \in R.$$

Proof.

Let R be an integral domain and $a \cdot b = a \cdot c$ and $a \neq 0$, for all $a, b, c \in R$.

We have $a \cdot b - a \cdot c = 0 \implies a \cdot (b - c) = 0$.

Therefore, since R is an integral domain and $a \neq 0$, $b - c = 0$. (R has no zero divisor).

Therefore, $b = c$. Hence, the cancellation law holds.

Converse Part: Assume that the cancellation law holds in a ring R.

Let $a \cdot b = 0$, for $a \neq 0$ and $b \in R$. We have

$$ab = 0 = a0$$

$$\implies b = 0.$$

Thus, $ab = 0$ in $R \implies a = 0$ or $b = 0$.

Therefore, R has no zero divisors.

Therefore, R is an integral domain.

4.2.14   Solved Problems

1. Prove that the set $Z_4 = \{[0], [1], [2], [3]\}$ is a commutative ring with respect to the binary operations addition modulo 4 $(+_4)$ and multiplication modulo 4 $(\times_4)$.

   Solution.

   Tables 4.1 and 4.2 are composition tables for addition modulo 4 $(+_4)$ and multiplication modulo 4 $(\times_4)$, respectively.

   From Tables 4.1 and 4.2, we get the following:

TABLE 4.1
Composition Table for $+_4$

$+_4$	[0]	[1]	[2]	[3]
[0]	[0]	[1]	[2]	[3]
[1]	[1]	[2]	[3]	[0]
[2]	[2]	[3]	[0]	[1]
[3]	[3]	[0]	[1]	[2]

TABLE 4.2
Composition Table for $\times_4$

$\times_4$	[0]	[1]	[2]	[3]
[0]	[0]	[0]	[0]	[0]
[1]	[0]	[1]	[2]	[3]
[2]	[0]	[2]	[0]	[2]
[3]	[0]	[3]	[2]	[1]

(i) All the entries in both the tables belong to Z_4.

Therefore, Z_4 is closed under the operations $+_4$ and $\times_4$.

(ii) In both the tables,

Entries in the first row = Entries in the first column

Entries in the second row = Entries in the second column

Entries in the third row = Entries in the third column

Entries in the fourth row = Entries in the fourth column.

Therefore, the operations $+_4$ and $\times_4$ are commutative in Z_4.

(iii) Also, for any $a, b, c \in Z_4$, we have

$$a +_4 (b +_4 c) = (a +_4 b) +_4 c$$

and $\quad a \times_4 (b \times_4 c) = (a \times_4 b) \times_4 c$

since $\quad 0 +_4 (1 +_4 2) = 0 +_4 3 = 3$

and $\quad (0 +_4 1) +_4 2 = (1 +_4 2) = 3.$

$\therefore \quad (0 +_4 1) +_4 2 = (0 +_4 1) +_4 2.$

Also, $\quad 1 \times_4 (2 \times_4 3) = 1 \times_4 2 = 2$

and $\quad (1 \times_4 2) \times_4 3 = 2 \times_4 3 = 2.$

$\therefore 1 \times_4 (2 \times_4 3) = (1 \times_4 2) \times_4 3.$

Thus, the operations $+_4$ and $\times_4$ are associative in Z_4.

(iv) [0] is the additive identity of Z_4, and [1] is the multiplicative identity of Z_4.

(v) Additive inverses of [0], [1], [2], [3] are, respectively, [0], [3], [2], [1].

Multiplicative inverses of the non-zero elements [1], [2], [3] are, respectively, [1], [2], [3].

(vi) If $a, b, c \in Z_4$, then

$$a \times_4 (b +_4 c) = (a \times_4 b) +_4 (a \times_4 c)$$

and $\quad (a +_4 b) \times_4 c = (a \times_4 c) +_4 (b \times_4 c).$

Thus, the operation $\times_4$ is distributive over $+_4$ in Z_4.

Hence, $(Z_4, +_4, \times_4)$ is a commutative ring with unity.

2. Show that $(\mathbb{Z}, +, \times)$ is an integral domain where $\mathbb{Z}$ is the set of all integers.

Solution.

We know the following:

(ℤ, +) is an abelian group.

(ℤ, ×) is a monoid.

The operation × is distributive over +.

(ℤ, ×) is commutative.

(ℤ, +, ×) is without zero divisors.

(ℤ, +, ×) is an integral domain.

3. Give an example of a ring which is not a field.

Solution.

The ring ℤ of all integers is an integral domain but not a field.

4.2.15   Problems for Practice

1. Discuss a ring and a field with suitable examples.

2. If $(R, +, \cdot)$ is a ring, then prove that $a \cdot 0 = 0$, for all $a \in R$, and 0 is the identity element in R under addition.

5

Lattices and Boolean Algebra

5.1 Introduction

In this chapter, we focus on partially ordered sets, lattices, Boolean algebra, and their properties. These structures are useful in set theory, algebra, sorting, and searching and in the construction of logical representation for computer science. The concept of the lattices is a special case of a partially ordered set. Boolean algebra is a special lattice.

5.2 Partial Ordering and Posets

Definition 5.2.1 Partial Order Relation: *A relation R on a non-empty set P is called a partial order, if R is reflexive, antisymmetric, and transitive. That is, if R satisfies*

 (i) xRx, *for all* $x \in P$ *(reflexive)*

 (ii) xRy *and* $yRx \implies x = y$, *for all* $x, y \in P$ *(antisymmetric)*

 (iii) xRy *and* $yRz \implies xRz$, *for all* $x, y, z \in P$ *(transitive).*

Example 5.2.2 *Let P be the set of all positive integers. Define the relation 'R' such that xRy holds if and only if $x \leq y$, for all $x, y \in P$. Clearly, "$\leq$" relation is reflexive, antisymmetric, and transitive. Hence, "$\leq$" relation on P is a partial order relation.*

Remark 5.2.3 *Usually, the partially ordered relation is denoted by "$\leq$".*

Definition 5.2.4 Partially Ordered Set or Poset: *A set P with the partial order relation "$\leq$" is called a partially ordered set or simply a poset. It is denoted by $\langle P, \leq \rangle$.*

Example 5.2.5 *Consider $P = \{$collection of all subsets of any set$\}$. Clearly, "$\subseteq$" relation (set inclusion) is a partially ordered relation on P.*

Definition 5.2.6 Totally Ordered Set: *Let $\langle P, \leq \rangle$ be a partially ordered set. If for every $a, b \in P$ we have either $a \leq b$ or $b \leq a$, then $\leq$ is called* *simple ordering* or *linear ordering* on P, and $\langle P, \leq \rangle$ is called a *totally ordered* or *simply ordered set* or a *chain*.

Example 5.2.7 *The poset $\langle \mathbb{Z}, \leq \rangle$ is totally ordered, since $a \leq b$ or $b \leq a$ whenever a and b are integers.*

Definition 5.2.8 Well-ordered Set: *A partially ordered set is called well-ordered if every non-empty subset of it has a least member.*

5.2.1   Representation of a Poset by Hasse Diagram

A partially ordered relation "$\leq$" on a set P can be represented by means of a diagram known as a Hasse diagram. In such a diagram, each element is represented by a small circle or a dot. The circle for an element x in P is drawn below the circle for y in P, if $x < y$, and a line is drawn between x and y, if y covers x. If $x < y$ but y does not cover x, then x and y are not connected directly by a single line.

Example 5.2.9 For example, let $P = \{1, 2, 3, 4\}$ and "$\leq$" be the relation "less than or equal to". Then, the Hasse diagram is shown below.

Hasse diagram of P

Example 5.2.10 Consider the set $X = \{2, 3, 6, 12, 24, 36\}$ and the relation "$\leq$" is defined as $x \leq y$ if and only if x divides y. The Hasse diagram of the poset $\langle X, \leq \rangle$ is shown below.

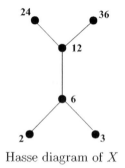

Hasse diagram of X

Note:

1. Hasse diagram is named after the twentieth-century German mathematician Helmut Hasse.

2. In a digraph, if we apply the following rules, then we get Hasse diagram.

   (i) Each vertex of a poset P must be related to itself. So, the arrows from vertex to itself are not necessary.

   (ii) If a vertex b appears above vertex a and if vertex a is connected to vertex b by an edge, then we have aRb; so, direction arrows are not necessary.

   (iii) If vertex c is above a and if c is connected to a by a sequence of edges, then we have aRc.

   (iv) The vertices are denoted by points rather than by circles.

Example 5.2.11 Let $A = \{a, b\}$. Let $B = P(A) = \{\{\phi\}, \{a\}, \{b\}, \{a, b\}\}$. Then, $\subseteq$ is a relation on A whose digraph and Hasse diagram are given in Figures 5.1 and 5.2.

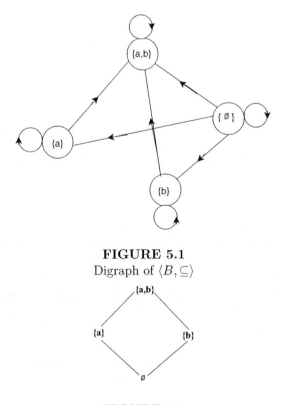

FIGURE 5.1
Digraph of $\langle B, \subseteq \rangle$

FIGURE 5.2
Hasse diagram of $\langle B, \subseteq \rangle$

5.2.2   Solved Problems

1. Show that the "greater than or equal" relation ($\geq$) is a partial ordering on the set of integers.

   Solution.

   (i) Since $a \geq a$ for every integer a, the relation $\geq$ is reflexive.

   (ii) If $a \geq b$ and $b \geq a$, then $a = b$. Hence, $\geq$ is antisymmetric.

   (iii) The relation $\geq$ is transitive since $a \geq b$ and $b \geq c$ imply that $a \geq c$.

   Hence, $\geq$ is a partial ordering on the set of integers, and $\langle \mathbb{Z}, \geq \rangle$ is a poset.

2. Show that the inclusion relation $\subseteq$ is a partial ordering on the power set of a set S.

   Solution.

   (i) Since $A \subseteq A$, whenever A is a subset of S, the relation $\subseteq$ is reflexive.

   (ii) Since $A \subseteq B$ and $b \subseteq A$ imply that $A = B$, the relation $\subseteq$ is antisymmetric.

   (iii) Since $A \subset B$ and $B \subset C$ imply that $A \subseteq C$, the relation $\subseteq$ is transitive.

   Therefore, the relation $\subseteq$ is a partial ordering on $P(S)$, and $\langle P(S), \subseteq \rangle$ is a poset.

3. Let R be a binary relation on the set of all positive integers such that $R = \{(a, b)/a = b^2\}$. Is R reflexive, symmetric, antisymmetric, transitive, an equivalence relation, or a partial order relation?

   Solution.

   (i) $R = \{(a, b)/a, b \text{ are positive integers and } a = b^2\}$.
        For R to be reflexive, we should have aRa, for all positive integers a. But aRa holds only when $a = a^2$ by hypothesis. Now, $a = a^2$ is not true for all positive integers. In fact, only when $a = 1$, we have $a = a^2$. Hence, R is not reflexive.

   (ii) For R to be symmetric, if aRb holds, then we should have bRa. But aRb implies $a = b^2$. But $a = b^2$ does not imply $b = a^2$ always for positive integers. For instance, $16 = 4^2$, but $4 \neq 16^2$. Hence, aRb does not imply bRa. Hence, R is not symmetric.

   (iii) For R to be antisymmetric, for positive integers a, b if aRb and bRa hold, then $a = b$. aRb implies $a = b^2$, and bRa implies $b = a^2$. Hence, if $a = b^2$ and $b = a^2$, then $a = b^2 = (a^2)^2 = a^4$, that is, $a^4 - a = 0$, that is, $a(a^3 - 1) = 0$. Since a is a positive integer, $a \neq 0$ so that $a^3 - 1 = 0$, that is, $a^3 = 1$ which implies $a = 1$. This means $b = a^2 = 1$. Hence, aRb and bRa imply $a = b = 1$. Hence, R is antisymmetric.

(iv) For R to be transitive, if aRb holds and bRc holds, then aRc should hold.

That is, aRb implies $a = b^2$, and bRc implies $b = c^2$, so that $a = b^2 = c^4$. Hence, aRc does not hold.

For example, $256 = 16^2$ and $16 = 4^2$ but $256 \neq 4^2$. Thus, R is not transitive.

(v) R is not an equivalence relation since an equivalence relation is reflexive, symmetric, and transitive.

(vi) R is not a partial order relation, since a partial ordering relation is reflexive, antisymmetric, and transitive.

4. Give examples of a relation which is both a partial ordering relation and an equivalence relation on a set.

Solution.
Equality and similarity of triangles are examples of a relation which is both a partial ordering relation and an equivalence relation.

5. Let S be a set. Determine whether there is a greatest element and a least element in the poset $\langle P(S), \subseteq \rangle$.

Solution.
The least element is the empty set since $\phi \subseteq T$ for any subset T of S. The set S is the greatest element in this poset. Hence $T \subseteq S$ whenever T is a subset of S.

6. Is there a greatest element and a least element in the poset $\langle \mathbb{Z}^+, | \rangle$?

Solution.
The integer 1 is the least element since 1 divides n whenever n is a positive integer. Since there is no integer that is divisible by all positive integers, there is no greatest element.

7. Let A be a given finite set and $P(A)$ its power set. Let $\subseteq$ be the inclusion relation on the elements of $P(A)$. Draw Hasse diagram of $\langle P(A), \subseteq \rangle$ for

    (i) $A = \{a\}$ **(ii)** $A = \{a, b\}$

    (iii) $A = \{a, b, c\}$ **(iv)** $A = \{a, b, c, d\}$.

Solution.

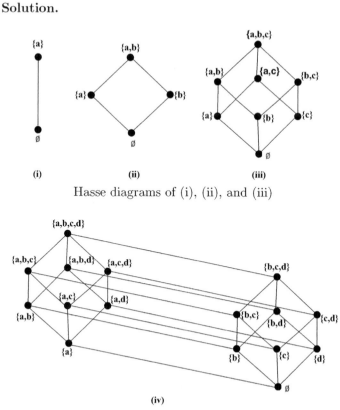

Hasse diagrams of (i), (ii), and (iii)

Hasse diagram of (iv)

8. Which elements of the poset $\langle\{2, 4, 5, 10, 12, 20, 25\}, |\rangle$ are maximal, and which of them are minimal?

Solution.

The Hasse diagram is shown in Figure 5.3.

From the Hasse diagram in Figure 5.3, this poset shows that the maximal elements are 12, 20, and 25 and the minimal elements are

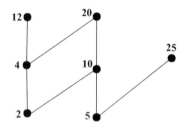

FIGURE 5.3

Hasse diagram of the given poset

2 and 5. As this example shows, a poset can have more than one maximal element and more than one minimal element.

9. Determine whether the posets represented by each of the Hasse diagrams in the following figure have a greatest element and a least element.

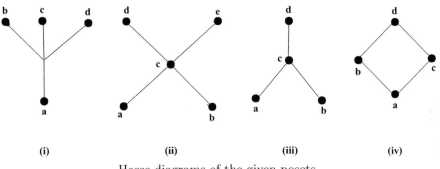

(i)   (ii)   (iii)   (iv)

Hasse diagrams of the given posets

Solution.

(i) The least element of the poset with Hasse diagram (i) is a. This poset has no greatest element.

(ii) The poset with Hasse diagram (ii) has neither a least nor a greatest element.

(iii) The poset with Hasse diagram (iii) has no least element. Its greatest element is d.

(iv) The poset with Hasse diagram (iv) has the least element a and greatest element d.

10. Draw the Hasse diagram of the set of partitions of 5.

Solution.

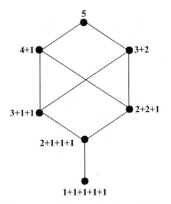

Hasse diagram of the set of partitions of 5

$$5 = 5$$
$$5 = 4 + 1$$
$$5 = 3 + 2$$
$$5 = 3 + 1 + 1$$
$$5 = 2 + 2 + 1$$
$$5 = 2 + 1 + 1 + 1$$
$$5 = 1 + 1 + 1 + 1 + 1.$$

5.2.3   Problems for Practice

1. Let R be the relation on the set of people such that xRy holds if x is older than y. Show that R is not a partial ordering.

2. Show that $\langle \mathbb{N}, \leq \rangle$ is a partially ordered set, where $\mathbb{N}$ is the set of all positive integers and $\leq$ is a relation defined by $m \leq n$ if and only if $n - m$ is a non-negative integer.

3. Show that there are only five distinct Hasse diagrams for partially ordered sets that contain three elements.

4. Give an example of a set X such that $\langle P(X), \subseteq \rangle$ is a totally ordered set.

5. Let S denote the set of all the partial ordering relations on a set P. Define a partial ordering relation on S, and interpret this relation in terms of the elements of P.

6. Let $X = \{1, 2, 3, 4, 6, 8, 12, 24\}$ and R be a division relation defined on X. Find the Hasse diagram of the poset $\langle X, R \rangle$.

7. Draw the Hasse diagrams of the following sets under the partial ordering relation "divides", and indicate those which are totally ordered:
   $\{2, 6, 24\}$, $\{3, 5, 15\}$, $\{1, 2, 3, 6, 12\}$, $\{2, 4, 8, 16\}$, $\{3, 9, 27, 54\}$.

8. If R is a partial ordering relation on a set X and $A \subseteq X$, show that $R \cap (A \times A)$ is a partial ordering relation on A.

9. Let $D_{30} = \{1, 2, 3, 5, 6, 10, 15, 30\}$, and let the relation R be divisor on D_{30}. Find

   (i) all the lower bounds of 10 and 15
   (ii) the greatest lower bound of 10 and 15
   (iii) all upper bounds of 10 and 15
   (iv) the least upper bound of 10 and 15.

   Also, draw the Hasse diagram.

10. Draw the Hasse diagram of $\langle X, \leq \rangle$, where $X = \{2, 4, 5, 10, 12, 20, 25\}$ and the relation $\leq$ be such that $x \leq y$ if x divides y.

Definition 5.2.12 Linearly ordered set or Chain: *A poset $\langle P, \leq \rangle$ is called a linearly ordered set or a chain if every pair of elements in a poset $\langle P, \leq \rangle$ are comparable.*

Example 5.2.13 *Let $\mathbb{Z}^+$ be the set of all positive integers. The usual relation "$\leq$" is a partial order relation on $\mathbb{Z}^+$, since any two integers in $\mathbb{Z}^+$ can be comparable with respect to the relation "$\leq$". Thus, $\langle \mathbb{Z}^+, \leq \rangle$ is a linearly ordered set.*

Definition 5.2.14 Upper bound and Lower bound: *Let S be any subset of a poset $\langle P, \leq \rangle$. An element $x \in P$ is called an upper bound of S if $y \leq x$, for all $y \in S$. An element $z \in P$ is called a lower bound of S if $z \leq y$, for all $y \in S$.*

Example 5.2.15 *Let $A = \{a, b, c\}$ be a given set and $\rho(A)$ be its power set. Let "$\subseteq$" be the relation on $\rho(A)$. Then clearly, $\langle \rho(A), \subseteq \rangle$ is a poset. For the subset $S = \{\{a, b\}, \{a\}, \{b\}, \{c\}\} \subseteq \rho(A)$, the upper bounds are $\{a, b\}$ and $\{a, b, c\}$, and its lower bound is ϕ.*

5.3 Lattices, Sublattices, Direct Product, Homomorphism of Lattices

Definition 5.3.1 Lattice: *A lattice is a poset $\langle L, \leq \rangle$ in which any subset $\{a, b\}$ consisting of two elements has a least upper bound and a greatest lower bound.*

We denote $LUB(\{a, b\})$ by $a \oplus b$ and call it as the join of a and b. Similarly, we denote $GLB(\{a, b\})$ by $a \star b$ and call it as the meet of a and b.

Example 5.3.2 *Let S be a set, and let $L = \rho(S)$. Let "$\subseteq$" (set inclusion) be the relation on L. Clearly, $\langle L, \subseteq \rangle$ is a lattice in which the meet and join are the same as the operations $\cap$ and $\cup$ on sets, respectively.*

That is, for any two elements $A, B \in \rho(S)$, $GLB(\{A, B\}) = A \cap B$ and $LUB(\{A, B\}) = A \cup B$.

Definition 5.3.3 Dual Lattices: *Let $\langle L, \leq \rangle$ be a poset, and let $\langle L, \geq \rangle$ be the dual poset (the symbol '$\geq$' used for the partial order is $\leq'$).*

If $\langle L, \leq \rangle$ is a lattice, we can show that $\langle L, \geq \rangle$ is also a lattice. In fact, for any $a, b \in L$, the $LUB(a, b)$ in $\langle L, \leq \rangle$ is equal to $GLB(a, b)$ in $\langle L, \geq \rangle$.

Similarly, the $GLB(a, b)$ in $\langle L, \leq \rangle$ is equal to $LUB(a, b)$ in $\langle L, \geq \rangle$.

Thus, the dual of $\langle L, \leq \rangle$ is $\langle L, \geq \rangle$ and vice-versa.

5.3.1 Properties of Lattices

In the following theorems, let $\langle L, \leq \rangle$ be a lattice.

Theorem 5.3.4 [Idempotent law] *For any* $a, b, c \in L$, *we have* $a \star a = a$ *and* $a \oplus a = a$.

Proof.
Let $a, b, c \in L$. Then by the definition of GLB of a and b, we have

$$a \star b \le a \qquad\qquad (5.1)$$

and if $a \le a$ and $a \le b$, then

$$a \le a \star b. \qquad\qquad (5.2)$$

Since $a \le a$, from (5.1) and (5.2), we have
$$a \star a \le a \quad \text{and} \quad a \le a \star a.$$
By the antisymmetric property, it follows that $a = a \star a$.
Similarly, we can prove that $a \oplus a = a$.

Theorem 5.3.5 [Associative law] *The operations of meet and join on* $\langle L, \le \rangle$ *are associative. That is, for any* $a, b, c \in L$, *we have the following:*

(i) $a \star (b \star c) = (a \star b) \star c$

(ii) $a \oplus (b \oplus c) = (a \oplus b) \oplus c.$

Proof.
To prove: $a \star (b \star c) = (a \star b) \star c.$
Let $a, b, c \in L$, Then by the definition, we have
$$(a \star b) \star c \le a \star b$$
and $\quad (a \star b) \star c \le c.$
By the definition of GLB of a and b, we have $a \star b \le a$ and $a \star b \le b$. Hence, by the transitive property of $\le$, we have
$$(a \star b) \star c \le a$$
and $\quad (a \star b) \star c \le b.$
Since $(a \star b) \star c \le b$ and $(a \star b) \star c \le c$, we see that $(a \star b) \star c$ is a lower bound for b and c. From the definition of $b \star c$, it follows that $(a \star b) \star c \le b \star c$.
Since $(a \star b) \star c \le a$ and $(a \star b) \star c \le b \star c$, from the definition of $a \star (b \star c)$, we have
$$(a \star b) \star c \le a \star (b \star c). \qquad\qquad (5.3)$$

Now, $a \star (b \star c) \le a$ and $a \star (b \star c) \le b \star c.$
Since $b \star c \le b$, by transitivity, we have $a \star (b \star c) \le b$.
Since $a \star (b \star c) \le a$ and $a \star (b \star c) \le b$, we have $a \star (b \star c) \le a \star b$.
Since $a \star (b \star c) \le b \star c \le$, we have

$$a \star (b \star c) \le (a \star b) \star c. \qquad\qquad (5.4)$$

From (5.3), (5.4) and by antisymmetric property, it follows that
$$a \star (b \star c) = (a \star b) \star c.$$
Similarly, we can prove that $a \oplus (b \oplus c) = (a \oplus b) \oplus c.$

Theorem 5.3.6 [Commutative law] *The operations of meet and join on* $\langle L, \leq \rangle$ *satisfy commutative property. That is, for any* $a, b \in L$, *we have the following:*

(i) $a \star b = b \star a$

(ii) $a \oplus b = b \oplus a$.

Proof.
Given: $a, b \in L$. Both $a \star b$ and $b \star a$ are GLB of a and b. By the uniqueness of GLB of a and b, we have $a \star b = b \star a$. Similarly, $a \oplus b = b \oplus a$ holds good.

Theorem 5.3.7 [Absorption law] *For any* $a, b \in L$, *we have the following:*

(i) $a \star (a \oplus b) = a$

(ii) $a \oplus (a \star b) = a$.

Proof.
Let $a, b \in L$. Then, $a \leq a$ and $a \leq a \oplus b$. So, $a \leq a \star (a \oplus b)$. On the other hand, $a \star (a \oplus b) \leq a$. By antisymmetric property of $\leq$, we have $a = a \star (a \oplus b)$.
Similarly, we have $a \oplus (a \star b) = a$, for all $a, b \in L$.

5.3.2 Theorems on Lattices

Theorem 5.3.8 *Let* $\langle L, \leq \rangle$ *be a lattice in which* $\star$ *and* $\oplus$ *denote the operations of meet and join respectively. For any* $a, b \in L$,
$$a \leq b \iff a \star b = a \iff a \oplus b = b.$$

Proof.
First, let us prove that $a \leq b \iff a \star b = a \iff a \oplus b = b$.
Let us assume that $a \leq b$, and also, we know that $a \leq a$.

$$\therefore \qquad\qquad a \leq a \star b. \qquad\qquad (5.5)$$

But, from the definition of $a \star b$, we have

$$a \star b \leq a. \qquad\qquad (5.6)$$

Hence, $a \leq b \implies a \star b = a$ [using (5.5) and (5.6)].
Next, assume that $a \star b = a$. But it is only possible if $a \leq b$.
That is, $a \star b = a \implies a \leq b$.
Combining these two results, we get
$$a \leq b \iff a \star b = a.$$
Similarly, we can prove that $a \leq b \iff a \oplus b = b$.
From $a \star b = a$, we have
$$b \oplus (a \star b) = b \oplus a = a \oplus b.$$
But $b \oplus (a \star b) = b$.
Hence, $a \oplus b = b$ follows that $a \star b = a$.

Theorem 5.3.9 *Let $\langle L, \leq \rangle$ be a lattice. For any $a, b \in L$, the following are equivalent:*

(i) $a \leq b$

(ii) $a \star b = a$

(iii) $a \oplus b = b$.

Proof.

First, consider (i) $\Longleftrightarrow$ (ii).

We have $a \leq a$. Assume $a \leq b$. Therefore, $a \leq a \star b$. By the definition of GLB, we have

$$a \star b \leq a.$$

Hence, by antisymmetric property, $a \star b = a$.

Assume that $a \star b = a$, but it is only possible if

$$a \leq b \Longrightarrow a \star b = a \Longrightarrow a \leq b.$$

Combining these two results, we have $a \leq b \Longleftrightarrow a \star b = a$.

Similarly, $a \leq b \Longleftrightarrow a \oplus b = b$.

Now, consider (ii) $\Longleftrightarrow$ (iii).

Assume $a \star b = a$, we have $b \oplus (a \star b) = b \oplus a = a \oplus b$, but by absorption, $b \oplus (a \star b) = b$.

$$\text{Hence, } a \oplus b = b.$$

By similar arguments, we can show that $a \star b = a$ follows from $a \oplus b = b$.

$$\text{(ii) } \Longleftrightarrow \text{ (iii)}$$

Hence, the theorem is proved.

Theorem 5.3.10 *Let $\langle L, \leq \rangle$ be a lattice. For any $a, b \in L$, the following inequalities hold:*

(1) Distributive Inequalities

(i) $a \oplus (b \star c) \leq (a \oplus b) \star (a \oplus c)$

(ii) $a \star (b \oplus c) \geq (a \star b) \oplus (a \star c)$.

(2) Modular Inequalities

(i) $a \leq c \Longleftrightarrow a \oplus (b \star c) \leq (a \oplus b) \star c$

(ii) $a \geq c \Longleftrightarrow a \star (b \oplus c) \geq (a \star b) \oplus c$.

Proof.

Since (ii) in (1) and (ii) in (2) are duals of (i) in (1) and (i) in (2) respectively, it is enough to prove (i) in (1) and (i) in (1) only.

Consider (i) in (1).

Let $a, b, c \in L$. Since $a \leq a \oplus b$ and $a \leq a \oplus c$, we have

$$a \leq [(a \oplus b) \star (a \oplus c)].$$

Since $b \star c \leq b \leq a \oplus b$ and $b \star c \leq c \leq a \oplus c$, we have

$$(b \star c) \leq (a \oplus b) \star (a \oplus c).$$

Therefore, $(a \oplus b) \star (a \oplus c)$ is an upper bound for a and $b \star c$, and hence

$$a \oplus (b \star c) \leq (a \oplus b) \star (a \oplus c).$$

Thus, (i) in (1) is proved.

The inequality (i) in (2) is a special case of (i) in (1).

If $a \leq c$, then $a \oplus c = c$, and from (i) in (1), we obtain

$$a \oplus (b \star c) \leq (a \oplus b) \star (a \oplus c) = (a \oplus b) \star c, \text{ which is inequality (i) in (2)}.$$

Hence, the theorem is proved.

Theorem 5.3.11 *In a lattice* $\langle L, \leq \rangle$, *for all* $a, b, c \in L$, *we have the following:*

(i) $(a \star b) \oplus (c \star d) \leq (a \oplus c) \star (b \oplus d)$

(ii) $(a \star b) \oplus (b \star c) \oplus (c \star a) \leq (a \oplus b) \star (b \oplus c) \star (c \oplus a)$.

Proof.

Let $a, b, c \in L$. Then

$$a \star b \leq a \quad \text{(or)} \quad b \leq a \oplus b. \tag{5.7}$$

$$a \star b \leq a \leq c \oplus a. \tag{5.8}$$

$$a \star b \leq b \leq b \oplus c. \tag{5.9}$$

Using (5.7), (5.8), and (5.9), we get

$$a \star b \leq (a \oplus b) \star (b \oplus c) \star (c \oplus a).$$

Similarly, $\quad b \star c \leq (a \oplus b) \star (b \oplus c) \star (c \oplus a)$,

$$c \star a \leq (a \oplus b) \star (b \oplus c) \star (c \oplus a).$$

This proves (ii).

We have $a \leq a \oplus c$ and $b \leq b \oplus d$.

We know that

$$c \leq a \oplus c. \tag{5.10}$$

$$d \leq b \oplus d. \tag{5.11}$$

Therefore, $\quad c \star d \leq (a \oplus c) \star (b \oplus d)$.

By (5.10) and (5.11), we have

$$(a \star b) \oplus (c \star d) \leq (a \oplus c) \star (b \oplus d).$$

This proves (i).

Theorem 5.3.12 *In a lattice* $\langle L, \leq \rangle$, *prove that for* $a, b, c \in L$,

(i) $(a \star b) \oplus (a \star c) \leq a \star (b \oplus (a \star c))$

(ii) $(a \oplus b) \star (a \oplus c) \geq a \oplus (b \star (a \oplus c))$.

Proof.

We know that $a \star b \leq a, \ a \star c \leq a$.

Therefore, $\qquad (a \star b) \oplus (a \star c) \leq a \oplus a = a. \tag{5.12}$

Also, $\qquad a \star b \leq b, \ a \star c \leq a \star c$

$\implies \qquad (a \star b) \oplus (a \star c) \leq b \oplus (a \star c). \tag{5.13}$

From (5.12) and (5.13), we have

$$(a \star b) \oplus (a \star c) \leq a \star (b \oplus (a \star c)).$$

This proves (i).

We know that $a \le a \oplus b; \ a \le a \oplus c$

$\implies$ $a = a \star a \le (a \oplus b) \star (a \oplus c).$ (5.14)

Further, $b \le a \oplus b; \ a \oplus c \le a \oplus c$

$\implies$ $b \star (a \oplus c) \le (a \oplus b) \star (a \oplus c).$ (5.15)

Using (5.14) and (5.15), we have
$$a \oplus (b \star (a \oplus c)) \le (a \oplus b) \star (a \oplus c).$$
This proves (ii).

Theorem 5.3.13 *In a lattice if $a \le b \le c$, show that*

(i) $a \oplus b = b \star c$

(ii) $(a \star b) \oplus (b \star c) = (a \oplus b) \star (a \oplus c) = b.$

Proof.

Let $a \le b \le c$.
$$a \le b \implies a \oplus b = b, \ a \star b = a.$$
$$b \le c \implies b \oplus d = c, \ b \star c = b.$$
$$a \le c \implies a \oplus c = c, \ a \star c = a.$$
Therefore, $a \oplus b = b = b \star c$, which is (i).
Now, $(a \star b) \oplus (b \star c) = a \oplus b = b$
$$(a \oplus b) \star (a \oplus c) = b \star c = b, \text{ which is (ii).}$$

5.3.3   Solved Problems

1.   Determine whether the posets represented by each of the Hasse diagrams are lattices.

   Solution.

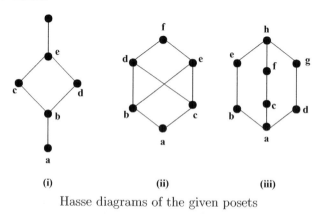

(i)   (ii)   (iii)

Hasse diagrams of the given posets

The posets represented by the Hasse diagrams in (i) and (iii) are both lattices because in each poset, every pair of elements has both a least upper bound and a greatest lower bound. On the other hand, the poset with the Hasse diagram shown in (ii) is not a lattice, since the elements b and c have no least upper bound.

It is to be noted that each of the elements d, e, and f is an upper bound, but none of these three elements precede the other two with respect to the ordering of this poset.

2. Is the poset $\langle \mathbb{Z}^+, | \rangle$ a lattice?

Solution.
Let a and b be two positive integers. The least upper bound and greatest lower bound of these two integers are the least common multiple and the greatest common divisor of these integers, respectively. Hence, it follows that this poset is a lattice.

3. Explain why the partially ordered sets of Figures 5.4 and 5.5 are not lattices.

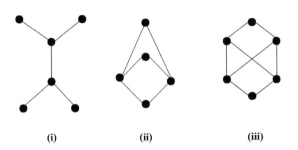

(i) (ii) (iii)

FIGURE 5.4
Hasse diagrams of the given posets

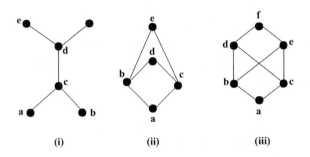

(i) (ii) (iii)

FIGURE 5.5
Hasse diagrams of the given posets

Solution. Given:

(i) does not represent a lattice since $e \oplus f$ does not exist.

(ii) does not represent a lattice since $b \oplus c$ does not exist.

(iii) does not represent a lattice because neither $d \oplus c$ nor $b \star c$ exists.

4. Let the sets $S_0, S_1, S_2, \ldots, S_7$ be given by

$$S_0 = \{a, b, c, d, e, f\}, \qquad S_1 = \{a, b, c, d, e\},$$
$$S_2 = \{a, b, c, d, f\}, \qquad S_3 = \{a, b, c, e\},$$
$$S_4 = \{a, b, c\}, \qquad S_5 = \{a, b\},$$
$$S_6 = \{a, c\}, \qquad S_7 = \{a\}.$$

Draw the diagram of $\langle L, \subseteq \rangle$ where $L = \{S_0, S_1, S_2, \ldots, S_7\}$.

Solution.

The diagram is shown below.

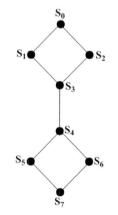

Hasse diagram of $\langle L, \subseteq \rangle$

5. Show that every non-empty subset of a lattice has a least upper bound and a greatest lower bound.

6. Show that every totally ordered set is a lattice.

7. Let $A = \{1, 2, 5, 10\}$ with the relation "divides". Draw the Hasse diagram.

Definition 5.3.14 Sublattice: *Let $\langle L, \star, \oplus \rangle$ be a lattice, and let $S \subseteq L$ be a subset of L. The algebra $\langle S, \star, \oplus \rangle$ is a sublattice of $\langle L, \star, \oplus \rangle$ if and only if S is closed under both operations $\star$ and $\oplus$.*

Example 5.3.15 *Let $\langle L, \leq \rangle$ be a lattice in which $L = \{a_1, a_2, \ldots, a_8\}$ and S_1, S_2, and S_3 be the sublattices of L given by $S_1 = \{a_1, a_2, a_4, a_6\}$, $S_2 = \{a_3, a_5, a_7, a_8\}$, and $S_3 = \{a_1, a_2, a_4, a_8\}$. The diagram of $\langle L, \leq \rangle$ is below.*

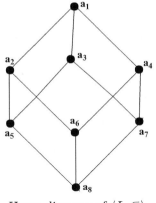

Hasse diagram of $\langle L, \subseteq \rangle$

Note that $\langle S_1, \leq \rangle$ and $\langle S_2, \leq \rangle$ are sublattices of $\langle L, \leq \rangle$, but $\langle S_3, \leq \rangle$ is not a sublattice since $a_2, a_4 \in S_3$ but $a_2 \star a_4 = a_6 \in S_3$. Also, note that $\langle S_3, \leq \rangle$ is a lattice.

Definition 5.3.16 Direct Product of Lattices: *Let $\langle L, \star, \oplus \rangle$ and $\langle S, \wedge, \vee \rangle$ be two lattices. The algebraic system $\langle L \times S, \cdot, + \rangle$ in which the binary operations "$+$" and "$\cdot$" on $L \times S$ are such that for any (a_1, b_1) and (a_2, b_2) in $L \times S$*
$$(a_1, b_1) \cdot (a_2, b_2) = (a_1 \star a_2, b_1 \wedge b_2)$$
$$(a_1, b_1) + (a_2, b_2) = (a_1 \oplus a_2, b_1 \vee b_2)$$
is called the direct product of the lattices $\langle L, \star, \oplus \rangle$ and $\langle S, \wedge, \vee \rangle$.

Definition 5.3.17 Lattice Homomorphism: *Let $\langle L, \star, \oplus \rangle$ and $\langle S, \wedge, \vee \rangle$ be two lattices. A mapping $g : L \longrightarrow S$ is called a lattice homomorphism from the lattice $\langle L, \star, \oplus \rangle$ to $\langle S, \wedge, \vee \rangle$ if for any $a, b \in L$,*
$$g(a \star b) = g(a) \wedge g(b) \quad and \quad g(a \oplus b) = g(a) \vee g(b).$$

Remark 5.3.18 *Observe that both the operations of meet and join are preserved. These may be mappings which preserve only one of the two operations. Such mappings are not lattice homomorphisms.*

Definition 5.3.19 Lattice Isomorphism: *If a homomorphism $g : L \longrightarrow S$ of two lattices $\langle L, \star, \oplus \rangle$ and $\langle S, \wedge, \vee \rangle$ is bijective, that is, one-to-one and onto, then g is called an isomorphism. If there exists an isomorphism between two lattices, then the lattices are called isomorphic.*

Definition 5.3.20 Lattice Endomorphism: *Let $\langle L, \star, \oplus \rangle$ be a lattice. A homomorphism $g : L \longrightarrow L$ is called an endomorphism.*

Definition 5.3.21 Lattice Automorphism: *Let $\langle L, \star, \oplus \rangle$ be a lattice. If $g : L \longrightarrow L$ is an isomorphism, then g is called an automorphism.*

Remark 5.3.22 *Let $\langle L, \star, \oplus \rangle$ be a lattice. If $g : L \longrightarrow L$ is an endomorphism, then the image set of g is a sublattice of L.*

Definition 5.3.23 Order-Preserving Mapping: *Let $\langle P, \leq \rangle$ and $\langle Q, \leq' \rangle$ be two partially ordered sets. A mapping $f : P \longrightarrow Q$ is said to be order-preserving relative to the ordering $\leq$ in P and $\leq'$ in Q if and only if for any $a, b \in P$ such that $a \leq b$, $f(a) \leq' f(b)$ in Q.*

Remark 5.3.24 *If $\langle P, \leq \rangle$ and $\langle Q, \leq' \rangle$ are lattices and $g : P \longrightarrow Q$ is a lattice homomorphism, then g is order-preserving.*

Definition 5.3.25 Order-isomorphic Partially Ordered Sets: *Two partially ordered sets $\langle P, \leq \rangle$ and $\langle Q, \leq' \rangle$ are called order-isomorphic if there exists a mapping $f : P \longrightarrow Q$ which is bijective and if both f and f^{-1} are order-preserving.*

5.3.4 Problem for Practice

1. Let $\langle L, \star, \oplus \rangle$ and $\langle S, \wedge, \vee \rangle$ be any two lattices with the partial orderings $\leq$ and $\leq'$ respectively. If g is a lattice homomorphism, then g preserves the partial ordering.

5.4 Special Lattices

Let $\langle L, \star, \oplus \rangle$ be a lattice and $S \subseteq L$ be a finite subset of L where $S = \{a_1, a_2, \ldots, a_n\}$. The greatest lower bound and the least upper bound of S can be expressed as

$$GLB \ \ S = \star_{i=1}^{n} a_i \quad \text{and} \quad LUB \ \ S = \oplus_{i=1}^{n} a_i$$

where $\quad \star_{i=1}^{2} a_i = a_1 \star a_2 \quad$ and $\quad \star_{i=1}^{k} a_i = \star_{i=1}^{k-1} (a_i \star a_k), \quad k = 3, 4, \ldots$

A similar representation can be given for $\oplus_{i=1}^{n}$. In lieu of the associative property of the operations $\star$ and $\oplus$, we can write

$$\star_{i=1}^{n} a_i = a_1 \star a_2 \star \cdots \star a_n$$

and $\qquad \oplus_{i=1}^{n} a_i = a_1 \oplus a_2 \oplus \cdots \oplus a_n.$

Definition 5.4.1 Complete Lattice: *A lattice is called complete if each of its non-empty subsets has a least upper bound and a greatest lower bound.*

Definition 5.4.2 Complement Element: *In a bounded lattice $\langle L, \star, \oplus, 0, 1 \rangle$, an element $b \in L$ is called a complement of an element $a \in L$ if*

$$a \star b = 0 \qquad and \qquad a \oplus b = 1.$$

Definition 5.4.3 Complemented Lattice: *A lattice $\langle L, \star, \oplus, 0, 1 \rangle$ is said to be a complemented lattice if every element of L has at least one complement.*

Definition 5.4.4 Distributive Lattice: *A lattice* $\langle L, \star, \oplus \rangle$ *is called a distributive lattice if for any* $a, b, c \in L$,

$$a \star (b \oplus c) = (a \star b) \oplus (a \star c)$$

$$a \oplus (b \star c) = (a \oplus b) \star (a \oplus c).$$

In other words, in a distributive lattice, the operations $\star$ *and* $\oplus$ *are distributed over each other.*

Definition 5.4.5 Modular Lattice: *A lattice* $\langle L, \wedge, \vee \rangle$ *is called modular if for all* $x, y, z \in L$,

$$x \le z \implies x \vee (y \wedge z) = (x \vee y) \wedge z \quad \text{(modular equations)}.$$

Remark 5.4.6 *We have (by modular inequality) if* $x \le z \implies x \vee (y \wedge z) = (x \vee y) \wedge z$ *holds in any lattice. Therefore, to show that a lattice L is modular, it is enough to show if*

$$x \le z \implies x \vee (y \wedge z) \ge (x \vee y) \wedge z \quad \text{holds in } L.$$

Theorem 5.4.7 *Every chain is a distributive lattice.*

Proof.

Let $\langle L, \le \rangle$ be a chain. Let $a, b, c \in L$. Consider the following possible cases:

(i) $a \le b$ or $a \le c$

(ii) $a \ge b$ and $a \ge c$.

We shall now show the distributive law

$$a \star (b \oplus c) = (a \star b) \oplus (a \star c).$$

In case (i), if $a \le b$ or $a \le c$, then we have

$$a \star b = a, a \oplus a = a, a \star c = c \quad \text{and}$$

$$\implies \qquad a \le b \oplus c. \tag{5.16}$$

Hence, $\qquad a \star (b \oplus c) = a$

and $\qquad (a \star b) \oplus (a \star c) = a \oplus a = a. \tag{5.17}$

From (5.16) and (5.17), we get

$$a \star (b \oplus c) = (a \star b) \oplus (a \star c).$$

In case (ii), if $a \ge b$ and $a \ge c$, then we have $a \star b = b$, $a \star c = c$ and $b \oplus c \le a$, so that

$$a \star (b \oplus c) = b \oplus c \tag{5.18}$$

and $\qquad (a \star b) \oplus (a \star c) = b \oplus c. \tag{5.19}$

From (5.18) and (5.19), we get

$$a \star (b \oplus c) = (a \star b) \oplus (a \star c).$$

Theorem 5.4.8 *Let* $\langle L, \star, \oplus \rangle$ *be a distributive lattice. For any* $a, b, c \in L$,

$$(a \star b = a \star c) \wedge (a \oplus b = a \oplus c) \Longrightarrow b = c.$$

Proof.

$$(a \star b) \oplus c = (a \star c) \oplus c = c. \tag{5.20}$$

$$\begin{aligned}
(a \star b) \oplus c &= (a \oplus c) \star (b \oplus c) \\
&= (a \oplus b) \star (b \oplus c) \\
&= b \oplus (a \star c) \\
&= b \oplus (a \star b) \\
&= b. \tag{5.21}
\end{aligned}$$

From (5.20) and (5.21), we have

$$b = c.$$

Theorem 5.4.9 *Every distributive lattice is modular.*

Proof.
Let $\langle L, \leq \rangle$ be a distributive lattice.
For all $a, b, c \in L$, we have

$$a \oplus (b \star c) = (a \oplus b) \star (a \oplus c).$$

Thus, if $a \leq c$, then $a \oplus c = c$ and

$$a \oplus (b \star c) = (a \oplus b) \star c.$$

Hence, if $a \leq c$, the modular equation is satisfied, and L is modular.

5.4.1   Solved Problems

1. Show that a chain of three or more elements is not complemented.

   Solution.
   In a chain, we have that any two elements are comparable.
   Let $0, x, 1$ be any three elements in a chain $\langle L, \leq \rangle$ with least element 0 and greatest element 1.
   We have $0 \leq x \leq 1$.
   Now, $0 \wedge x = 0$ and $0 \vee x = x$.
   Similarly, $x \wedge 1 = x$ and $x \vee 1 = 1$.
   Therefore, x does not have any complement.
   Hence, any chain with three or more elements is not complemented.

2. Find all sublattices of $\langle D_{30}, | \rangle$ where $|$ is the divisor relation.

   Solution.
   The Hasse diagram of $\langle D_{30}, | \rangle$ is shown in Figure 5.6.

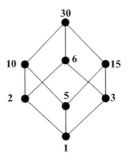

FIGURE 5.6

Hasse diagram of $\langle D_{30}, | \rangle$

Therefore, the sublattices are

$$D_6 = \{1, 2, 3, 6\}$$
$$D_{10} = \{1, 2, 5, 10\}$$
$$D_{15} = \{1, 3, 5, 15\}$$
$$S_1 = \{5, 10, 15, 30\}$$
$$S_2 = \{3, 5, 15, 30\}, \text{ etc. are lattices.}$$

In general, if $m|n$, then D_m is a sublattice of D_n, and D_{km} is also a sublattice of D_n.

3. Show that the lattices given by the diagrams are not distributive.

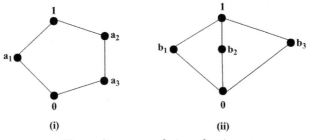

Hasse diagrams of given lattices

Solution.

In lattice (i),

$$a_3 \star (a_1 \oplus a_2) = a_3 \star 1 = a_3 = (a_3 \star a_1) \oplus (a_3 \star a_2)$$
$$a_1 \star (a_2 \oplus a_3) = 0 = (a_1 \star a_2) \oplus (a_1 \star a_3)$$
but $\quad a_2 \star (a_1 \oplus a_3) = a_2 \star 1 = a_2$
$$(a_2 \star a_1) \oplus (a_2 \star a_3) = 0 \oplus a_3 = a_3.$$

Hence, the lattice (i) is not distributive.

In lattice (ii),

$b_1 \star (b_2 \oplus b_3) = b_1$ while $(b_1 \star b_2) \oplus (b_1 \star b_3) = 0$ which shows that the lattice is not distributive.

4. If D_n denotes the lattice of all the divisors of the integer n, draw the Hasse diagrams of D_{10}, D_{15}, D_{32}, and D_{45}.

Solution.
The Hasse diagrams are shown below.

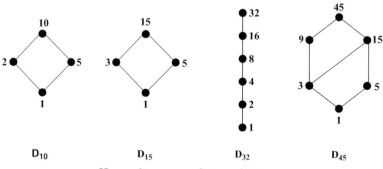

Hasse diagrams of given lattices

5. Prove that in a distributive lattice, the complement of an element is unique.

Solution.
Let a be an element with two distinct complements b and c. Then
$$a \star b = 0 \quad \text{and} \quad a \star c = 0$$
$$\implies \quad a \star b = a \star c.$$
Also, $a \oplus b = 1 \quad \text{and} \quad a \oplus c = 1$
$$\implies \quad a \oplus b = a \oplus c.$$
By a theorem, we have $b = c$.

6. Let L be a complemented, distributive lattice. Then for $a, b \in L$, show the following are equivalent:
   (i) $a \leq b$
   (ii) $a \star b' = 0$
   (iii) $a' \oplus b = 1$
   (iv) $b' \leq a'$

   where $'$ denotes corresponding complement.

                                    or
   Show that the following hold in a distributive and complemented lattice L:
   $$a \leq b \iff a \star b' = 0 \iff a' \oplus b = 1 \iff b' \leq a' \text{ for } a, b \in L.$$

Solution.

$$a \le b \Longrightarrow a \oplus b = b$$
$$\Longrightarrow (a \oplus b) \star b' = 0 \quad \text{since} \quad b \star b' = 0$$
$$\Longrightarrow (a \star b') \oplus (b \star b') = 0$$
$$\Longrightarrow a \star b' = 0 \qquad \text{since} \quad b \star b' = 0.$$

Hence (i) $\Longrightarrow$ (ii).

$$a \star b' = 0 \Longrightarrow (a \star b)' = 1$$
$$\Longrightarrow a' \oplus (b')' = 1$$
$$\Longrightarrow a' \oplus b = 1.$$

Hence (ii) $\Longrightarrow$ (iii).

$$a' \oplus b = 1 \Longrightarrow (a' \oplus b) \star b' = b'$$
$$\Longrightarrow (a' \star b') \oplus (b \star b') = b' \quad \text{(using distributive law)}$$
$$\Longrightarrow a' \star b' = b' \qquad \text{since} \quad b \star b' = 0$$
$$\Longrightarrow b' \le a'.$$

Hence (iii) $\Longrightarrow$ (iv).

7. Let $\langle L, \wedge, \vee \rangle$ be a distributive lattice and $a, b, c \in L$. If $a \wedge b = a \wedge c$ and $a \vee b = a \vee c$, then $b = c$.

or

Show that the cancellation laws are valid in a distributive lattice.

Solution.
Let $\langle L, \wedge, \vee \rangle$ be a distributive lattice and $a, b, c \in L$, such that $a \wedge b = a \wedge c$ and $a \vee b = a \vee c$. Now,

$$(a \wedge b) \vee c = (a \vee c) \wedge (b \vee c) \quad \text{(since } L \text{ is distributive)}$$
$$= (a \vee b) \wedge (b \vee c)$$
$$= (b \vee a) \wedge (b \vee c)$$
$$= b \vee (a \wedge c)$$
$$= b \vee (a \wedge b)$$
$$= b$$

and $\qquad (a \wedge b) \vee c = (a \wedge c) \vee c = c.$

Thus, $\qquad b = (a \wedge b) \vee c = c$, so that
$$a \wedge b = a \wedge c \text{ and } a \vee b = a \vee c \Longrightarrow b = c.$$

That is, the cancellation law is valid in a distributive lattice.

8. Show that the direct product of any two distributive lattices is a distributive lattice.

Solution.

Let L_1 and L_2 be two distributive lattices. Let $x, y, z \in L_1 \times L_2$, the direct product (lattice) of L_1 and L_2. Then, $x = (a_1, a_2)$, $y = (b_1, b_2)$, and $z = (c_1, c_2)$ for some $a_1, b_1, c_1 \in L_1$ and $a_2, b_2, c_2 \in L_2$. Now,

$$x \vee (y \wedge z)$$
$$= (a_1, a_2) \vee ((b_1, b_2) \wedge (c_1, c_2))$$
$$= (a_1, a_2) \vee (b_1 \wedge c_1, b_2 \wedge c_2)$$
$$= (a_1 \vee (b_1 \wedge c_1), a_2 \vee (b_2 \wedge c_2))$$
$$= ((a_1 \vee b_1) \wedge (a_1 \vee c_1), (a_2 \vee b_2) \wedge (a_2 \vee c_2))$$
$$\text{(since } L_1 \text{ and } L_2 \text{ are distributive lattices)}$$
$$= ((a_1 \vee b_1), (a_2 \vee b_2)) \wedge ((a_1 \vee c_1), (a_2 \vee c_2))$$
$$= ((a_1, a_2) \vee (b_1, b_2)) \wedge ((a_1, a_2) \vee (c_1, c_2))$$
$$= (x \vee y) \wedge (x \vee z).$$

Hence, for all $x, y, z \in L_1 \times L_2$, $x \vee (y \wedge z) = (x \vee z) \wedge (x \vee z)$.

Therefore, if L_1 and L_2 are distributive lattices, then the direct product $L_1 \times L_2$ is also a distributive lattice.

9. Prove that the lattice is modular.

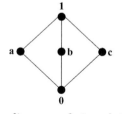

Hasse diagram of given lattice

Solution.

The elements a, b, and c are symmetric in the lattice. It is enough to prove for any one of a, b, c.

We have the cases $a < 1$ and $0 < a$.

Case (i): Let $a < 1$.

Let $x_1 = a$ and $x_3 = 1$. Then
$$x_1 \vee (x_2 \wedge x_3) = a \vee (x_1 \wedge 1) = a \vee x_2$$
and
$$(x_1 \vee x_2) \wedge x_3 = (a \vee x_2) \wedge 1 = a \vee x_2.$$
Hence, $x_1 \vee (x_2 \wedge x_3) = (x_1 \vee x_2) \wedge x_3$.

Case (ii): Let $0 < a$.

Let $x_1 = 0$ and $x_3 = a$. Then
$$x_1 \vee (x_2 \wedge x_3) = 0 \vee (x_2 \wedge a) = x_2 \wedge a$$
and
$$(x_1 \vee x_2) \wedge x_3 = (0 \vee x_2) \wedge a = x_2 \wedge a.$$
Hence, $x_1 \vee (x_2 \wedge x_3) = (x_1 \vee x_2) \wedge x_3$.

Therefore, the above lattice is modular.

5.4.2 Problems for Practice

1. Find the complements, if they exist, of the elements a, b, c of the lattice, whose Hasse diagram is given below. Can the lattice be complemented?

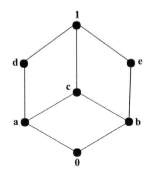

2. Give an example for a distributive and complemented lattice.

3. Examine whether the lattice given in the following Hasse diagram is distributive or not.

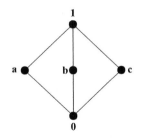

4. In a distributive complemented lattice, show that the following are equivalent:

   (i) $a \leq b$
   (ii) $a \wedge \bar{b} = 0$
   (iii) $\bar{a} \vee b = 1$
   (iv) $\bar{b} \leq \bar{a}$.

5. Let $\langle L, \leq, \vee, \wedge \rangle$ be a distributive lattice and $a, b \in L$ if $a \wedge b = a \wedge c$ and $a \vee b = a \vee c$. Then, show that $b = c$.

6. Define a lattice. Give a suitable example.

7. In a complemented and distributive lattice, prove that the complement of each element is unique.

8. State modular inequality of lattices.

9. Show that cancellation laws are valid in a distributive lattice.

5.5   Boolean Algebra

Definition 5.5.1 Boolean Algebra: *A Boolean algebra is a complemented distributive lattice.*

A Boolean algebra will generally be denoted by $\langle B, \star, \oplus, ', 0, 1 \rangle$, and it satisfies the following properties in which a, b, and c denote any element of the set B.

1. $\langle B, \star, \oplus, ', 0, 1 \rangle$ is a lattice and satisfies the following:
   - (i) $a \star a = a$
   - (ii) $a \star b = b \star a$
   - (iii) $(a \star b) \star c = a \star (b \star c)$
   - (iv) $a \star (a \oplus b) = a$
   - (v) $a \oplus a = a$
   - (vi) $a \oplus b = b \oplus a$
   - (vii) $(a \oplus b) \oplus c = a \oplus (b \oplus c)$
   - (viii) $a \oplus (a \star b) = a$.

2. $\langle B, \star, \oplus \rangle$ is a distributive lattice and satisfies the following:
   - (i) $a \star (b \oplus c) = (a \star b) \oplus (a \star c)$
   - (ii) $a \oplus (b \star c) = (a \oplus b) \star (a \oplus c)$
   - (iii) $(a \star b) \oplus (b \star c) \oplus (c \star a) = (a \oplus b) \star (b \oplus c) \star (c \oplus a)$
   - (iv) $(a \star b) = (a \star c)$ and $(a \oplus b) = (a \oplus c) \Longrightarrow b = c$.

3. $\langle B, \star, \oplus, ', 0, 1 \rangle$ is a bounded lattice and satisfies the following:
   - (i) $0 \leq a \leq 1$
   - (ii) $a \star 0 = 0$
   - (iii) $a \star 1 = a$
   - (iv) $a \oplus 0 = a$
   - (v) $a \oplus 1 = 1$.

4. $\langle B, \star, \oplus, ', 0, 1 \rangle$ is a complemented lattice in which the complement of any element $a \in B$ is denoted by $a' \in B$ and satisfies the following:
   - (i) $a \star a' = 0$
   - (ii) $a \oplus a' = 1$
   - (iii) $0' = 1$
   - (iv) $1' = 0$
   - (v) $(a \star b)' = a' \oplus b'$
   - (vi) $a \oplus b' = a' \star b'$.

5. There exists a partial ordering $\leq$ on B such that

    (i) $a \star b = GLB\{a, b\}$

    (ii) $a \oplus b = LUB\{a, b\}$

    (iii) $a \leq b \iff a \star b = a \iff a \oplus b = b$

    (iv) $a \leq b \iff a \star b' = 0 \iff b' \leq a' \iff a' \oplus b = 1$.

Example 5.5.2 *Let* $A = \{a, b, c\}$ *and consider the lattice* $\langle P(A), \cap, \cup \rangle$ *as shown below.*

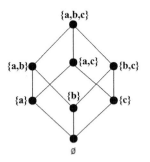

Hasse diagram of $\langle P(A), \cap, \cup \rangle$

Clearly, $\langle P(A), \cap, \cup \rangle$ *is a Boolean algebra.*

Example 5.5.3 *Let* $B = \{0, 1\}$ *be a set. The operations* $\star, \oplus, '$ *on* B *are defined in the table below.*

Tables showing Operations of $\star, \oplus, '$ on B

$\star$	0	1	$\oplus$	0	1	x	x'
0	0	0	0	0	1	0	1
1	0	1	1	1	1	1	0

Clearly, $\langle B, \star, \oplus, ', 0, 1 \rangle$ *is a Boolean algebra.*

Definition 5.5.4 Sub-Boolean Algebra: *Let* $\langle B, \star, \oplus, ', 0, 1 \rangle$ *be a Boolean algebra and* $S \subseteq B$. *If* S *contains the elements* 0 *and* 1 *and is closed under the operations* $\star$, $\oplus$, $'$, *then* $\langle S, \star, \oplus, ', 0, 1 \rangle$ *is called a sub-Boolean algebra.*

Remark 5.5.5 *A sub-Boolean algebra of a Boolean algebra is itself a Boolean algebra.*

Remark 5.5.6 *A subset of a Boolean algebra can be a Boolean algebra. However, it may not be a sub-Boolean algebra because it may not close with respect to the operations in the Boolean algebra.*

Definition 5.5.7 Direct Product of Boolean Algebra: *Let* $\langle B_1, \star_1, \oplus_1,{}', 0_1, 1_1 \rangle$ *and* $\langle B_2, \star_2, \oplus_2,{}'', 0_2, 1_2 \rangle$ *be any two Boolean algebras. The direct product of the two Boolean algebras is defined to be a Boolean algebra that is given by* $\langle B_1 \times B_2, \star_3, \oplus_3,{}''', 0_3, 1_3 \rangle$ *in which the following operations are defined for any* $(a_1, b_1), (a_2, b_2) \in B_1 \times B_2$ *as*

$$(a_1, b_1) \star_3 (a_2, b_2) = ((a_1 \star_1 a_2), (b_1 \star_2 b_2))$$
$$(a_1, b_1) \oplus_3 (a_2, b_2) = ((a_1 \oplus_1 a_2), (b_1 \oplus_2 b_2))$$
$$(a_1, b_1)''' = (a_1', b_1'')$$
$$0_3 = (0_1, 0_2) \quad and \quad 1_3 = (1_1, 1_2).$$

Definition 5.5.8 Join-irreducible: *Let* $\langle L, \star, \oplus \rangle$ *be a lattice. An element* $a \in L$ *is called join-irreducible if it cannot be expressed as the join of two distinct elements of L.*

    In other words, $a \in L$ *is join-irreducible, if for any* $a_1, a_2 \in L$,
$$a = a_1 \oplus a_2 \implies (a = a_1) \star (a = a_2).$$

Definition 5.5.9 Boolean Homomorphism: *Let* $\langle B, \star, \oplus,{}', 0, 1 \rangle$ *and* $\langle P(A), \cup, \cap,{}^c, \alpha, \beta \rangle$ *be any two Boolean algebras, where A is a set. Then, a mapping* $f : B \longrightarrow P(A)$ *is called a Boolean homomorphism, if for any* $a, b \in B$,

$$f(a \star b) = f(a) \cap f(b)$$
$$f(a \oplus b) = f(a) \cup f(b)$$
$$f(a') = [f(a)]^c$$
$$f(0) = \alpha$$
$$f(1) = \beta.$$

Remark 5.5.10 *The binary operations* $\star$ *and* $\oplus$ *are preserved under Boolean homomorphism.*

Remark 5.5.11 *Let* $\langle L, \star, \oplus, \leq \rangle$ *and* $\langle S, \wedge, \vee, \leq' \rangle$ *be two Boolean algebras. Then, a mapping* $g : L \implies S$ *is called an order homomorphism, then*
$$a \leq b \implies g(a) \leq' g(b), \text{ for all } a, b \in L.$$

Theorem 5.5.12 *In a Boolean algebra, De Morgan's laws hold.*

Proof.
Let $\langle L, \star, \oplus,{}^-, 0, 1 \rangle$ be a Boolean algebra. Then, L is a complemented and distributive lattice.
De Morgan's laws are
$$\overline{a \oplus b} = \bar{a} \star \bar{b}, \quad \overline{a \star b} = \bar{a} \oplus \bar{b}, \text{ for all } \bar{a}, a, b \in L.$$
    Assume that $a, b \in L$. There exist elements $\bar{a}, \bar{b} \in L$ such that
$$a \oplus \bar{a} = 1, \quad a \star \bar{a} = 0, \quad b \oplus \bar{b} = 1, \quad b \star \bar{b} = 0.$$

(i) Claim: $a \overline{\oplus} b = \bar{a} \star \bar{b}$.

$$(a \oplus b) \oplus (\bar{a} \star \bar{b}) = [(a \oplus b) \oplus \bar{a}] \star [(a \oplus b) \oplus \bar{b}]$$
$$= [a \oplus \star a \oplus b] \star [a \oplus b \oplus \bar{b}]$$
$$= [a \oplus b] \star [a \oplus a]$$
$$= 1 \star 1 = 1.$$

$$(a \oplus b) \star (\bar{a} \star \bar{b}) = [(a \oplus b) \star \bar{a}] \star [(a \oplus b) \star \bar{b}]$$
$$= [(a \star \bar{a}) \oplus (b \star \bar{a})] \star [(a \star \bar{b}) \oplus (b \star \bar{b})]$$
$$= [0 \oplus (b \star \bar{a})] \star [(a \star \bar{b}) \oplus 0]$$
$$= (b \star \bar{a}) \star (a \star \bar{b})$$
$$= b \star (\bar{a} \star a) \star b = \bar{b} \star 0 \star \bar{b} = 0.$$

Hence, claim (i) is proved.

(ii) Claim: $a \overline{\star} b = \bar{a} \oplus \bar{b}$.

$$(a \star b) \oplus (\bar{a} \oplus \bar{b}) = [(a \star b) \oplus \bar{a}] \oplus [(a \star b) \oplus \bar{b}]$$
$$= [(a \oplus \bar{a}) \star (b \oplus \bar{a})] \oplus [(a \oplus \bar{b}) \star (b \oplus \bar{b})]$$
$$= [1 \star (b \oplus \bar{a})] \oplus [(a \oplus \bar{b}) \star 1]$$
$$= (b \oplus \bar{a}) \oplus (a \oplus \bar{b})$$
$$= b \oplus (\bar{a} \oplus a) \oplus \bar{b}$$
$$= b \oplus 1 \oplus \bar{b} = b \oplus \bar{b} = 1.$$

$$(a \star b) \star (\bar{a} \oplus \bar{b}) = [(a \star b) \star \bar{a}] \oplus [(a \star b) \star \bar{b}]$$
$$= (a \star \bar{a} \star b) \oplus (a \star b \star \bar{b})$$
$$= (0 \star b) \oplus (a \star 0)$$
$$= 0 \star 0 = 0.$$

Hence, claim (ii) is proved.
Therefore, De Morgan's laws are proved.

Theorem 5.5.13 *In a Boolean algebra* $\langle L, \star, \oplus \rangle$, *the complement* $\bar{a}$ *of any element* $a \in L$ *is unique.*

Proof.

Let $a \in L$ have two complements $b, c \in L$.

By definition, we have $a \star b = 0$, $a \oplus b = 1$, $a \star c = 0$, $a \oplus c = 1$. Then, we have

$$b = b \star 1$$
$$= b \star (a \oplus c)$$
$$= (b \star a) \oplus (b \star c)$$
$$= 0 \oplus (b \star c)$$
$$= b \star c \tag{5.22}$$

and
$$c = c \star 1$$
$$= c \star (a \oplus b)$$
$$= (c \star a) \oplus (c \star b)$$
$$= 0 \oplus (c \star b)$$
$$= c \star b$$
$$= b \star c. \tag{5.23}$$

From (5.22) and (5.23), we have $b = c$.

Therefore, every element of L has a unique complement.

5.5.1 Solved Problems

1. Show that $\langle P(A), \cup, \cap, \subseteq \rangle$ is a Boolean algebra, where A is any set.

   Solution.

   We know that $\langle P(A), \cup, \cap, \subseteq \rangle$ is a lattice.

   For any $X, Y, Z \in P(A)$,
   $$X \cap (Y \cup Z) = (X \cap Y) \cup (X \cap Z)$$
   $$X \cup (Y \cap Z) = (X \cup Y) \cap (X \cup Z).$$
   Also, for all $X \in P(A)$, there exists a subset $\bar{X}$ of A such that
   $$X \cup \bar{X} = A, \quad X \cap \bar{X} = \{\ \} = \phi.$$
   Zero element of $P(A)$ is $\{\ \} = $ least element.

   The greatest element of $P(A)$ is A.

   Therefore, $\langle P(A), \cup, \cap, \subseteq \rangle$ is a Boolean algebra.

2. Show that in any Boolean algebra,
   $$(a + b)(a' + c) = ac + a'b + bc.$$

   Solution.

   Let $\langle B, +, \cdot, \, ' \rangle$ be a Boolean algebra. Let $a, b, c \in B$.
   $$(a + b)(a' + c) = (a + b)a' + (a + b)c$$
   $$= aa' + ba' + ac + bc$$
   $$= 0 + a'b + ac + bc$$
   $$= ac + a'b + bc.$$

3. In any Boolean algebra, show that $a = b$ if and only if $a\bar{b} + \bar{a}b = 0$.

   Solution.

   Let $\langle B, +\cdot, \, ^-, 0, 1 \rangle$ be any Boolean algebra. Let $a, b \in B$ and $a = b$.

   To show that: $a\bar{b} + \bar{a}b = 0$.
   $$a \cdot \bar{b} + \bar{a} \cdot b = a\bar{a} + \bar{a}a = 0 + 0 = 0.$$

   Now, let $a\bar{b} + \bar{a}b = 0$, for all $a, b \in B$. Then
   $$a\bar{b} + \bar{a}b = 0$$

$$\implies \qquad a\bar{a} + a\bar{b} + \bar{a}b + b\bar{b} = 0$$
$$\implies \qquad a(\bar{a} + \bar{b}) + b(\bar{a} + \bar{b}) = 0$$
$$\implies \qquad (a + b)(\bar{a} + \bar{b}) = 0$$
$$\implies \qquad (a + b)\bar{a}\bar{b} = 0$$
$$\implies \qquad ab = a + b$$
$$\implies \qquad GLB\{a, b\} = LUB\{a, b\}$$
$$\implies \qquad a = b.$$

4. Simplify (i) $(a \star b)' \oplus (a \oplus b)'$
    (ii) $(a' \star b' \star c) \oplus (a \star b' \star c) \oplus (a \star b' \star c').$

Solution.

(i)
$$(a \star b)' \oplus (a \oplus b)' = (a' \oplus b') \oplus (a' \star b')$$
$$= (a' \oplus b' \oplus a') \star (a' \oplus b' \oplus b')$$
$$= (a' \oplus b') \star (a' \star b')$$
$$= a' \star b'.$$

(ii)
$$(a' \star b' \star c) \oplus (a \star b' \star c) \oplus (a \star b' \star c')$$
$$= (a' \oplus a) \star (b' \star c)$$
$$= 1 \star (b' \star c) = b' \star c.$$

5. Let a, b, c be any elements in a Boolean algebra B. Prove that

$$\text{(i) } a \star a = a \quad \text{(ii) } a \oplus a = a.$$

Solution.

(i) To prove: $a \star a = a$.

Let	$a = a \star 1$	(by identity law)
	$= a \star (a \oplus a')$	(by complement law)
	$= a \star a \oplus a \star a'$	(by distributive law)
	$= (a \star a) \oplus 0$	(by complement law)
	$= a \star a.$	(by identity law)

(ii) To prove: $a \oplus a = a$.

Let	$a = a \oplus 0$	(by identity law)
	$= a \oplus (a \star a')$	(by complement law)
	$= (a \oplus a) \star (a \oplus a')$	(by distributive law)
	$= (a \oplus a) \star 1$	(by complement law)
	$= a \oplus a.$	(by identity law)

6. Let a, b, c be any elements in a Boolean algebra B. Show that

$$\text{(i)} \quad a \oplus a = 1 \quad \text{(ii)} \quad a \star 0 = 0.$$

Solution.

(i) To prove: $a \oplus 1 = 1$.

Let	$a \oplus 1 = (a \oplus 1) \star 1$	(by identity law)
	$= (a \oplus 1) \star (a \oplus a')$	(by complement law)
	$= a \oplus (a \star a')$	(by distributive law)
	$= a \oplus (a' \star 1)$	(by commutative law)
	$= a \oplus a'$	(by identity law)
	$= 1.$	(by complement law)

(ii) To prove: $a \star 0 = 0$.

Let	$a \star 0 = (a \star 0) \oplus 0$	(by identity law)
	$= (a \star 0) \oplus (a \star a')$	(by complement law)
	$= a \star (0 \oplus a')$	(by distributive law)
	$= a \star (a' \oplus 0)$	(by commutative law)
	$= a \star a'$	(by identity law)
	$= 0.$	(by complement law)

7. Prove that $a \oplus (a' \star b) = a \oplus b$.
   Solution.
   $$a \oplus (a' \star b) = (a \oplus a') \star (a \oplus b) = 1 \star (a \oplus b) = a \oplus b.$$

8. Prove that $a \star (a' \oplus b) = a \star b$.
   Solution.
   $$a \star (a' \oplus b) = (a \star a') \oplus (a \star b) = 0 \oplus (a \star b) = a \star b.$$

9. Prove that $(a \star b) \oplus (a \star b') = a$.
   Solution.
   $$(a \star b) \oplus (a \star b') = a \star (b \oplus b') = a \star 1 = a.$$

10. In any Boolean algebra, $\langle B, \cdot, +, ', 0, 1 \rangle$, show that
    $$(a + b')(b + c')(c + a') = (a' + b)(b' + c)(c' + a).$$
    Solution.

$$\begin{aligned}
(a + b')(b + c')(c + a') &= (a + b' + 0)(b + c' + 0)(c + a' + 0) \\
&= (a + b' + cc')(b + c' + aa')(c + a' + bb') \\
&= (a + b' + c)(a + b' + c')(b + c' + a) \\
&\quad (b + c' + a')(c + a' + b)(c + a' + b') \\
&= [(a' + b + c)(a' + b + c')]
\end{aligned}$$

$$[(b' + c + a)(b' + c + a')]$$
$$[(c' + a + b)(c' + a + b')]$$
$$= (a' + b + cc')(b' + c + aa')(c' + a + bb')$$
$$= (a' + b + 0)(b' + c + 0)(c' + a + 0)$$
$$= (a' + b)(b' + c)(c' + a).$$

11. In any Boolean algebra, $\langle B, \cdot, +, \,', 0, 1 \rangle$, show that
$$a = 0 \Longleftrightarrow ab' + a'b = b.$$

Solution.
If $a = 0$, then it directly follows that
$$ab' + a'b = 0 + 1b = 0 + b = b.$$

Suppose $\qquad\qquad\qquad b = ab' + a'b. \qquad\qquad\qquad (5.24)$

Therefore, $\qquad\qquad 0 = b'b = b'(ab' + a'b) = ab' + 0 = ab'.$

Using De Morgan's law, from (5.24) we obtain $b'(a' + b)(a + b')$.
Therefore,

$$0 = ab' = a(a' + b)(a + b')$$
$$= (aa' + ab)(a + b')$$
$$= (0 + ab)(a + b')$$
$$= ab(a + b')$$
$$= aba + abb' = ab + 0 = ab.$$

Therefore, $\qquad 0 = ab = ab'.$
Therefore, $\qquad 0 = ab + ab' = a(b + b') = a1 = a.$
$$\text{Hence, } a = 0.$$

5.5.2 Problems for Practice

1. What values of the Boolean variables x and y satisfy $xy = x + y$?

2. Show that De Morgan's laws hold in a Boolean algebra. That is, show that for all x and y, $\overline{x \vee y} = \overline{x} \wedge \overline{y}$ and $\overline{x \wedge y} = \overline{x} \vee \overline{y}$.

3. Does a Boolean algebra contain six elements? Justify your answer.

4. If $P(S)$ is the power set of a non-empty set S, prove that $\langle P(S), \cup, \cap, {}^c, \phi, S \rangle$ is a Boolean algebra.

5. Prove that in a Boolean algebra, $(a \vee b)' = a' \wedge b'$.

6. Give an example of a two-element Boolean algebra.

7. Write the Boolean algebra whose Hasse diagram is a chain.

8. Is there a Boolean algebra with five elements? Justify your answer.

9. Show that a lattice homomorphism on a Boolean algebra which preserves 0 and 1 is a Boolean homomorphism.

10. Prove that a lattice with five elements is not a Boolean algebra.

11. Show that in a Boolean algebra, $a \oplus (a' \star b) = a \oplus b$.

12. Show that in a Boolean algebra, $a \star (a' \oplus b) = a \star b$.

13. Show that in a Boolean algebra, $(a \star b) \oplus (a \star b') = a$.

14. Show that in a Boolean algebra, $(a \star b \star c) \oplus (a \star b) = a \star b$.

15. Show that in a Boolean algebra, $a \leq b \Longrightarrow a + bc = b(a + c)$.

16. Simplify the Boolean expression: $(a \star c) \oplus c \oplus [(b \oplus b') \star a]$.

17. Simplify the Boolean expression: $(1 \star a) \oplus (0 \star a')$.

Bibliography

G. Balaji, *Discrete Mathematics*, G. Balaji Publishers, Chennai, India, 2017.

R. K. Bisht and H. S. Dhami, *Discrete Mathematics*, Oxford University Press, New Delhi, India, 2015.

Susanna S. Epp, *Discrete Mathematics with Applications*, Fourth Edition, Brooks/Cole, Cengage Learning, U.S.A., 2011.

Rowan Garnier and John Taylor, *Discrete Mathematics: Proofs, Structures and Applications*, Third Edition, CRC Press, Boca Raton, FL, 2010.

Seymour Lipschutz and Marc Lipson, *Discrete Mathematics*, Third Edition, Schaum's Outlines, Tata McGraw-Hill Company, New Delhi, India, 2007.

Kenneth H. Rosen, *Discrete Mathematics and its Applications*, Seventh Edition, McGraw-Hill Education, New York, U.S.A., 2007.

J. P. Tremblay and R. Manohar, *Discrete Mathematical Structures with Applications to Computer Science*, Tata McGraw-Hill Publishing Company Limited, New Delhi, India, 2008.

Index

A

Abelian group, 174, 184
Absorption law, 233
Adjacency matrix, 149
Adjacent edges, 137
Adjacent vertices, 137
Algebraic structure, 174
Algebra, 173
Algebraic system, 173
Associative property, 173

B

Biconditional, 4
Binary operation, 173
Bipartite graph, 143
Block, 158
Boolean algebra, 248
Boolean homomorphism, 250
Bounded variable, 23

C

Cancellation property, 174
Canonical form, 9
Cayley's representation
      theorem, 196
Chain, 231
Characteristic equation, 91
Chinese postman problem, 163
Circuit, 157
Circular path, 157
Closed walk, 157
Closure property, 173
Combination, 80
Combinatorics, 39
Commutative group, 184
Commutative ring, 219
Commutativity, 173

D

Complemented lattice, 240
Complete bipartite graph, 144
Complete graph, 142
Complete lattice, 240
Component, 158
Conditional statement, 3
Congruence relation, 175
Conjunction, 2
Connected graph, 158
Complement element, 240
Connectivity, 156
Contingency, 6
Contradiction, 6
Cosets, 197
Cycle, 157
Cycle graph, 142
Cyclic group, 193
Cyclic permutation, 209

D

Degree of a vertex, 138
Directed graph, 138
Direct product of Boolean
      algebra, 250
Direct product of lattices, 239
Direct proof, 22
Disconnected graph, 158
Disjoint cycles, 209
Disjunction, 3
Distributive lattice, 241
Distributive properties, 174
Duality law, 7
Dual lattices, 231

E

Elementary cycle, 157
Elementary path, 157